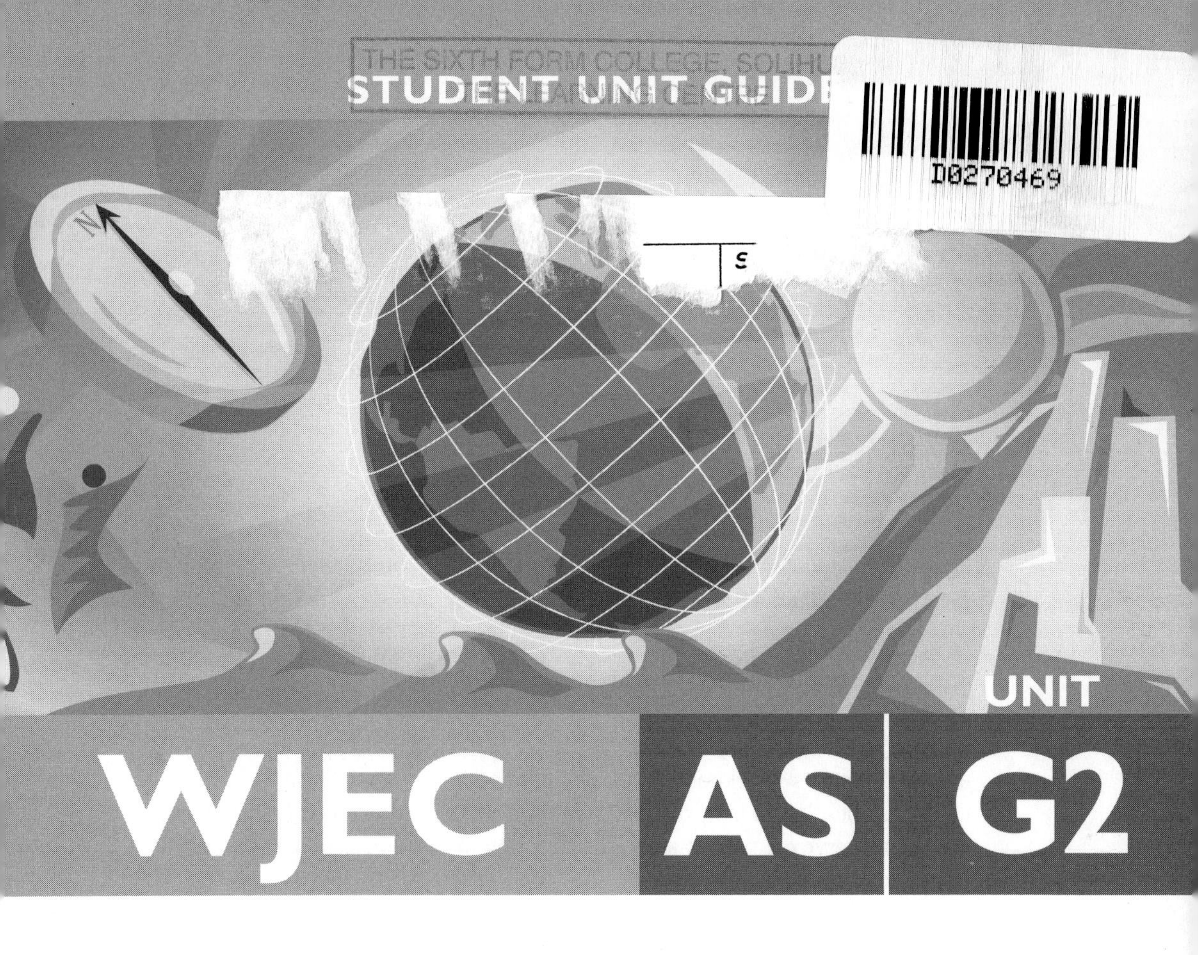

WJEC AS G2

Geography

Changing Human Environments

David Burtenshaw & Sue Warn

Philip Allan Updates, an imprint of Hodder Education, an Hachette UK company, Market Place, Deddington, Oxfordshire OX15 0SE

Orders

Bookpoint Ltd, 130 Milton Park, Abingdon, Oxfordshire OX14 4SB
tel: 01235 827720
fax: 01235 400454
e-mail: uk.orders@bookpoint.co.uk
Lines are open 9.00 a.m.–5.00 p.m., Monday to Saturday, with a 24-hour message answering service. You can also order through the Philip Allan Updates website: www.philipallan.co.uk

ISBN 978-1-4441-1083-8

First printed 2010
Impression number 5 4 3 2 1
Year 2014 2013 2012 2011 2010

This guide has been written specifically to support students preparing for the WJEC AS Geography Unit G2 examination. The content has been neither approved nor endorsed by WJEC and remains the sole responsibility of the authors.

Typeset by Pantek Arts Ltd, Maidstone, Kent.
Printed by MPG Books, Bodmin

Hachette UK's policy is to use papers that are natural, renewable and recyclable products and made from wood grown in sustainable forests. The logging and manufacturing processes are expected to conform to the environmental regulations of the country of origin.

P01668

Contents

Introduction

■■■

Content Guidance

■■■

Questions and Answers

Introduction

About this guide

The purpose of this guide is to help you understand what is required to do well in **Unit G2: Changing Human Environments**. The full contents of the specification are available on the WJEC website **www.wjec.co.uk**. This paper may be taken in Welsh. The comments all apply no matter what your language medium.

To help you with your studies this guide is divided into three sections. The **Introduction** explains the structure of the guide and how to approach the topics that you will be studying. It also provides some guidance on how to approach the Unit examination. The **Content Guidance** section gives you the bare bones of the specification for the two themes that you have to study. This section contains several useful diagrams and maps, which you may be able to use in the examination. The third part of this assessment is a techniques and field studies question related to the subject matter of the two themes. Unit G2 is an assessment which relies on you writing mini-essays in continuous prose. To help you with developing your geographical writing skills the **Questions & Answers** section includes examples of the types of question that you will see in the examination. Sample answers to Questions 1 and 2 at grade A and C are provided together with examiner's comments on how to tackle each question and how to improve answers to questions. The third question expects you to demonstrate your ability to interpret resources and to be able to write about your own investigation in human geography.

The questions

You have to answer all three questions, which are made up of three parts each. There is no choice of questions. The first question will test your understanding of **Theme 1: Investigating population change**. The second question will test your understanding of **Theme 2: Investigating settlement change in MEDCs**. The format for both of these questions is identical; there is an introductory part (a) (5 marks) that will use a resource such as a map, photograph, diagram, cartoon, or a short piece of prose to test your knowledge and understanding and interpretation of the resource. Part (b) (10 marks) may ask you a further question on the same area of study so that you can demonstrate your understanding of the topic. Sometimes this question may be on a separate area of the specification. Part (c) (10 marks) takes a look at an issue or an aspect of change that will enable you to apply your knowledge.

The format for the third question is slightly different. Part (a) (7 marks) will involve the interpretation of a resource in the form of a diagram, map, statistics, a table, photographs (including oblique aerial and satellite images) and/or a short passage of text. Part (b) (8 marks) will ask you to show how you can use various methods to interpret and analyse geographical data about changing human geography. Finally, Part (c) (10 marks) will focus on your own field or secondary data investigations in human geography.

Timing

The examination is 1 hour and 30 minutes, in which time you must answer all three questions. The topics appear in the same order as in the specification. You therefore have approximately 30 minutes per question, which is the equivalent of almost a mark to be gained every minute. You may answer questions in any order. The table below gives you some idea of the timings that you should ideally try to follow.

	Activity	**Time in minutes**
	Read all three questions. Decide on the order of questions to be answered. (The question order below assumes that you answer questions in order.)	1
Q1	Study the first resource carefully and answer the 5-mark question Part (a).	5
	Read Part (b) and plan your answer.	2
	Write your answer (probably about one side of prose plus any maps or diagrams).	10
	Read Part (c) and plan your answer.	2
	Write your answer (probably about one side of prose plus any maps or diagrams).	10
Q2	Study the second resource carefully and answer the 5 mark question Part (a).	5
	Read Part (b) and plan your answer.	2
	Write your answer (probably about one side of prose plus any maps or diagrams).	10
	Read Part (c) and plan your answer.	2
	Write your answer (probably about one side of prose plus any maps or diagrams).	10
Q3	Study the resource(s) for Question 3.	2
	Plan response to part (a).	1
	Write your answer to (a).	6
	Read and plan response to part (b).	1
	Write answer to (b).	8
	Read and plan response to part (c).	1
	Write one-side answer to (c).	9
	Check work.	3
	Total time	**90 minutes**

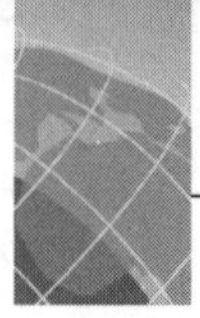

Quality of written communication

There are no specifically allocated marks for the quality of your writing. However, you should try to use correct punctuation and grammar, structure answers into a logical sequence with a brief introduction and conclusion, and use appropriate geographical terminology. If you are dyslexic or have other special needs that inhibit your ability to write you should make sure that your school has sought special consideration on your behalf.

Geographical terms

It is a good idea to collect your own list of key terms (your own geographical dictionary) as you come across them in your studies. It is also useful to build up your own memory bank of simple, effective maps and diagrams. Practise drawing your maps and diagrams because, ideally, you need to be able to spend no more than two minutes drawing a map or diagram. Finally, you will need your own examples to illustrate the points that you make in your answers. You will come across examples and diagrams in this book; remember that you can impress by the use of different, relevant examples that you have found rather than the standard ones in your textbooks.

Where to find good examples

Always try to find examples from your own studies in addition to the ones provided by teachers. *Topic Eye 2010* has a volume on Changing Cities. *Geography Review* and *Geographical* can be good sources of original examples. Quality newspapers, e.g. *Times, Guardian, Independent* and *Daily Telegraph*, are another source that provide good examples. But even with these papers you need to be aware of bias. There is good geography constantly in the media so make use of it.

Managing questions in a minute

All questions have two major components. There are **command words** such as outline, discuss, evaluate, assess. Secondly, there is the **subject matter**. Use highlighter pens in different colours to emphasise the commands and the topic.

How answers are marked

Markers always assess the *overall* quality of an answer against the marks scheme for that paper. The holistic mark might draw upon qualities from several levels and award a mark that best fits the combination of qualities in the answer. In addition all marks are also based on assessment objectives (AOs) which enable marks to be awarded for knowledge, application of that knowledge, and skills. These may be quality of language skills, or the skills of drawing or interpreting maps, photos and diagrams.

Content Guidance

Unit G2: Changing Human Environments consists of two themes:

- Investigating population change
- Investigating settlement change in MEDCs

Unit G2 focuses on population change at a global scale which is impacting on people at all levels from the global to the local. It also examines changes to settlements in the more developed world that you see going on around you. This unit provides opportunities for field studies to support the principles, theories and issues that you are studying.

Investigating population change comprises six sections:

1.1 What is demographic change?

1.2 How and why do populations change naturally?

1.3 What is the role of migration in population change?

1.4 What are the issues of the migration of refugees and asylum seekers?

1.5 What are the causes and impacts of changing gender structures?

1.6 What are the demographic challenges facing countries?

Investigating settlement change in MEDCs comprises six further sections:

2.1 What are the distinctive features of settlements?

2.2 How does the social and cultural structure of settlements vary and why?

2.3 What are the issues of the inner city?

2.4 What are the issues being faced in the CBD?

2.5 How is the rural–urban fringe changing and why?

2.6 How are rural settlements changing and why?

Theme 1: Investigating population change

The emphasis here is on change and therefore the concepts, ideas, issues and problems will all reflect the theme of change and adapting to change. Population is dynamic for it is always changing as people are born, die and migrate: it is the flow of people through time.

1.1 What is demographic change?

Population growth and changing population are among our greatest challenges. The six billionth person was born in 1999. **Systems theory** provides a framework for studying population because it enables us to see any population as dynamic. **Demography** is the study of population's vital statistics. This section looks at the tools you need to demonstrate your understanding of the fundamental principles. Population change can be measured at a variety of scales. The world's population doubled between 1960 and 2000. However, the rate of increase varies (Figure 1). The UK's population will take 433 years to double compared with 27 years in Africa. Since 1950 Africa and Asia have contributed almost 75% of the world's population growth whereas Europe has contributed only 11%. As a result, the distribution of the world's population is changing. World population growth is a global challenge, especially as it is occurring in those countries least equipped to cope. However, do bear in mind that the art of forecasting is very imprecise and is based mainly on projecting existing trends.

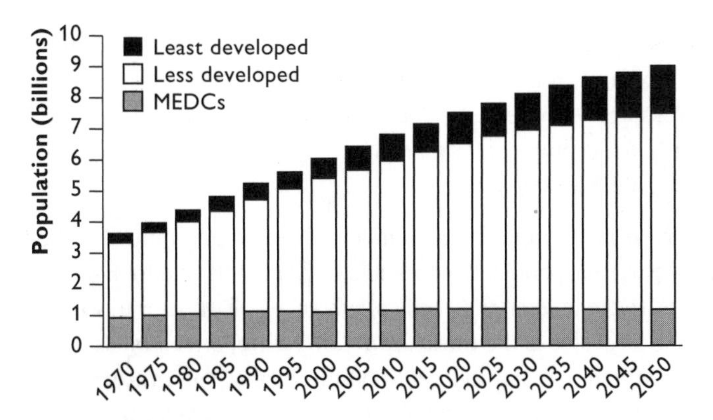

Figure 1 Population growth by development category, 1970–2050

There are now 22 countries with over 50 million people (there were 9 in 1950). In 2005, China had approximately 1304 million inhabitants and India 1104 million. The countries with the highest increases at the turn of the century were Islamic. By 2050 it is projected that the population of Yemen will rise by 240%, Saudi Arabia by 101% and Kuwait by 172%.

On 27 August 2009 it was announced that the estimated population of the UK was 61 million. Key patterns that you need to be aware of are **density** and **change** (natural and migration). You should be able to offer reasons for these patterns. In the UK population is growing at 0.7% per year compared with 0.3% in the 1990s. Women are having more babies but there has also been a 20% rise in women of child-bearing age as a result of migration. Consequently, the children of women born outside the UK contributed 56% of natural increase in 2008. Migration has contributed less to the absolute growth in numbers because many migrants, especially from Eastern Europe, are returning home and this is combined with the emigration of UK citizens.

Population and demographic data

Population data are normally collected by ten-yearly national censuses. These give the absolute numbers at that time, either where people are, or where they normally are. Some data are sampled; normally 10% of households is the sample size. Even censuses may lack accuracy because people do not complete forms. These days sampling between censuses by national statistical offices and by organisations such as the UN does give us more detail for intercensal periods and this type of data has been used in Tables 1 and 2.

Key terms

Natural increase is the excess of births over deaths. **Crude birth rate** is the ratio of the number of births per annum to the total population. **Total fertility rate** is generally measured as the number of births to a woman over her lifetime. This can be refined further to examine the number of births to a specified age group per 1000 women. In MEDCs this rate needs to be 2.1 children per woman aged 15–45 to maintain the level of population. As Table 1 shows, in Italy the rate was 1.3 in 2005 while in the UK it was 1.7. In Central African states in 2005 the average rate was 5.6 children per woman and in Islamic countries it was 4.1. Birth rates and fertility rates can be driven by:

- Culture, for example religious beliefs and dogma.
- Political pressures to increase or decrease the population.
- Social norms, for example later marriage.
- Events such as the end of a war.

Crude death rate is the ratio of the number of deaths per annum to the total population, expressed per 1000 people. It is possible to refine this and look at the **age-specific mortality rate.** Hygiene, diet and medical advances all affect the rise and fall of these rates. One of the key rates is the **infant mortality rate,** which measures the number of deaths of children under 1 year old per 1000 live births. The **maternal mortality rate** measures the deaths of women in childbirth per 100,000 live births. Both these last two measures are frequently used to measure the development of a country.

Life expectancy is the average number of years a person is expected to live. It can be refined by age groups and gender. It was 78 for children born in the UK in 2005 (MEDC average 76) whereas in Botswana it was 35 (LEDC average 65) and for all Southern African countries it was 50.

Migration is the movement of people from one administrative area to another, which results in the change of permanent residence. **Migration balance** is the excess of in-migration over out-migration, or vice versa. **In-migration** is the flow of people over a given time period (year or decade) into a country/area, on either a temporary or a permanent basis. **Out-migration** is the opposite flow.

Net population change is the change in population in a country over a period of time when both natural change and migration change have been taken into account. It is normally expressed as a number or a percentage. Figure 2 shows data for 1995–2000.

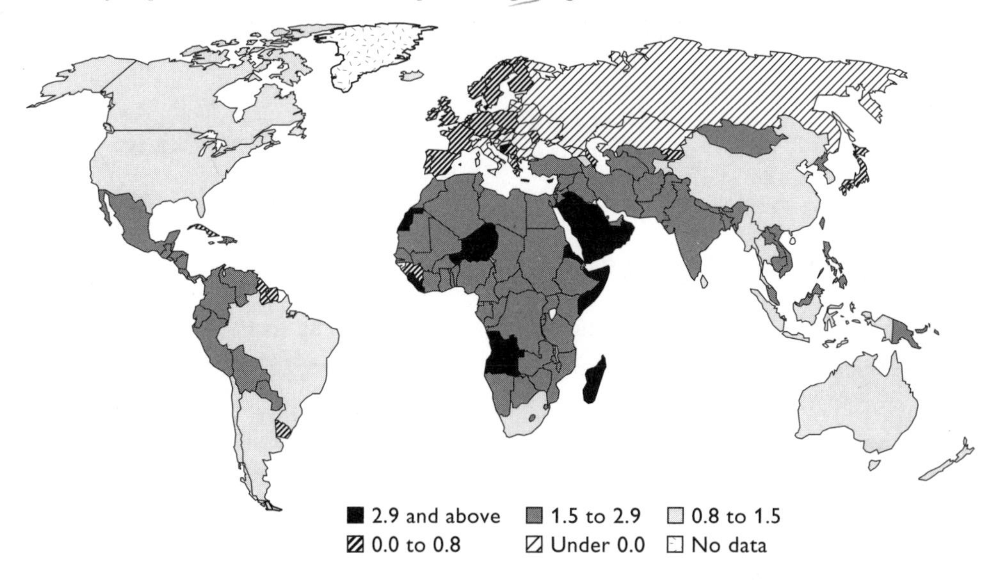

***Figure 2** Annual average rate of population change, 1995–2000 (%)*

Displaying demographic and population data

Geographers and demographers portray population in a variety of ways. Each of the measures above can be mapped at any scale from the global to the local (city).

Density maps show the number of people per unit of land area (persons per hectare). This is shown as a **choropleth map** (shading or colouring areas of different densities). This technique is used to illustrate most of the data above. Density maps ignore the ability of the land/country to support the population living within its boundaries. **Physiological density** measures the number of people supported per hectare of arable land, thus measuring a country's ability to support itself from within its own boundaries.

Distribution is the location of people portrayed by dots or proportional symbols. It is relatively unsatisfactory because symbols overlap, but it can prove useful at a local scale where there is more room to portray data.

Topological maps portray the global population challenge more graphically. The area of each country (or administrative area) is in proportion to its total population or other total being measured (see Figure 3).

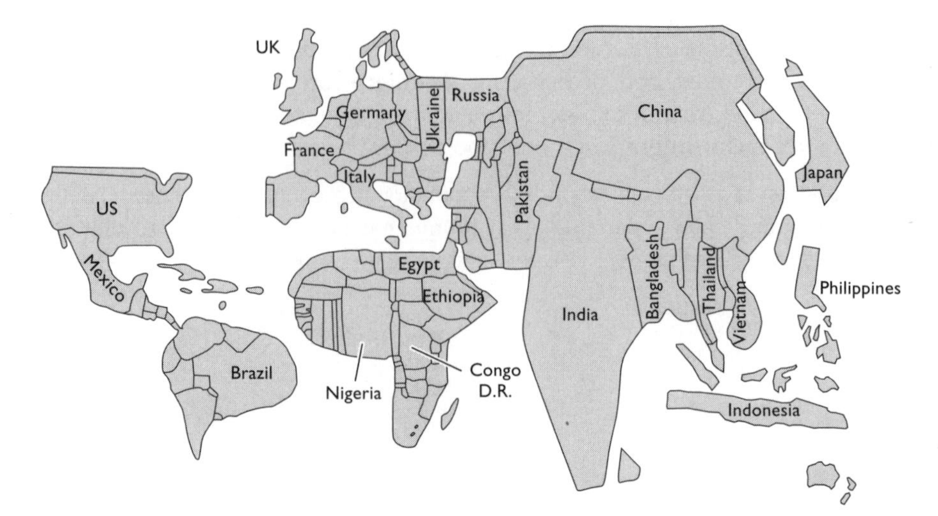

Figure 3 Topological map of the world's population

Migration may be portrayed using choropleth and dot map techniques but it is best portrayed on **flow maps** showing the direction and scale of movements by the width of the flow line.

The **age–sex pyramid** classically portrays gender and age. Pyramids are an indicator of development. They are used to study the demographic characteristics of individual groups in an area. Figures 4a, b and c show some pyramids for the residents of Singapore. In order to make pyramids comparable, all data should be expressed in percentages of the total population. However, many sources use absolute numbers, which make comparisons more difficult.

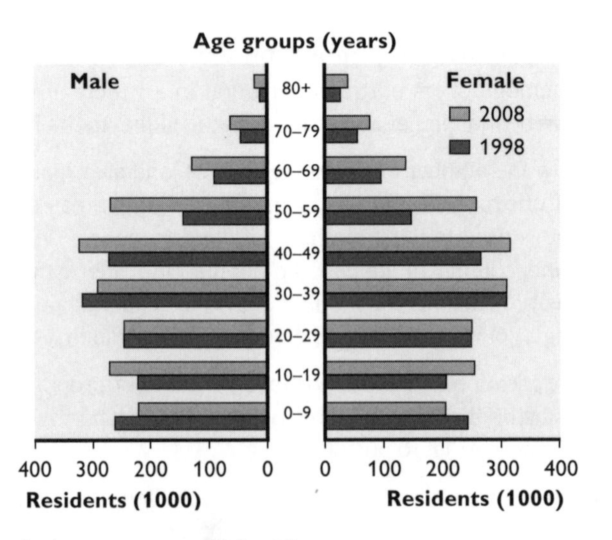

Figure 4a Population age pyramid for Singapore

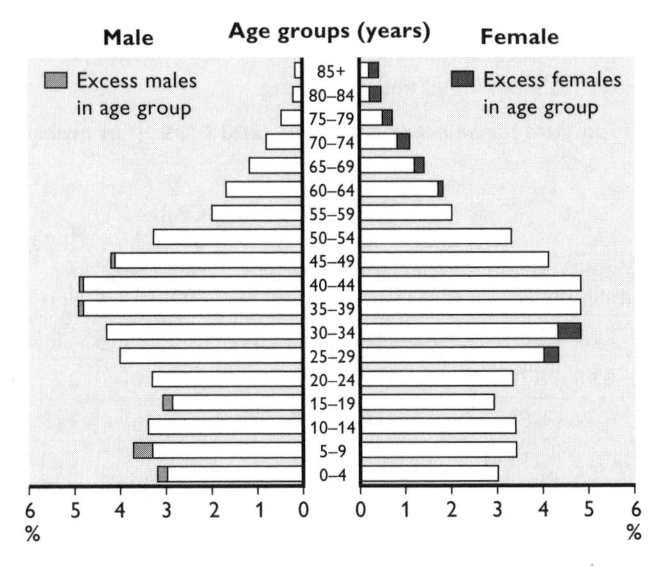

***Figure 4b* Chinese in Singapore (2000) as % of Chinese population (2.5 million)**

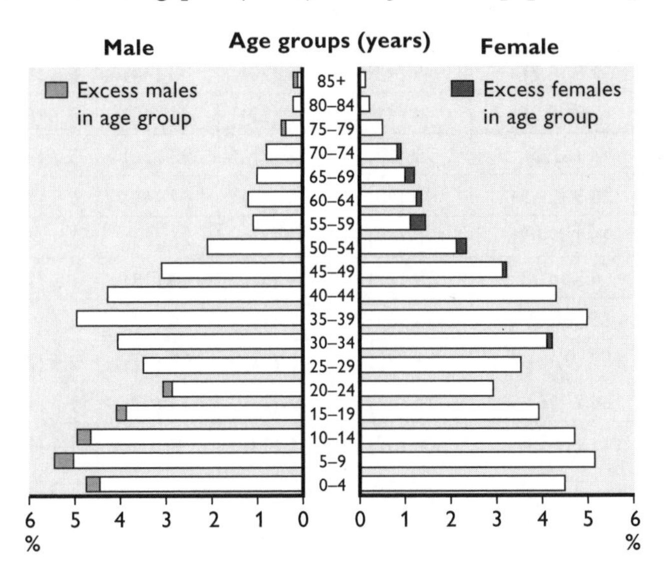

***Figure 4c* Malays in Singapore (2000) as % of Malay population (0.45 million)**

Graphs can also be used to show trends. **Line graphs** are used to show changed rates of growth for many of the demographic indicators. **Bar graphs** are used to illustrate time series for a set of places.

Global population change

Table 1 provides a range of indicators for selected countries. These countries provide useful data to enable you to answer questions. The data illustrate issues such as the

impact of war (Afghanistan) on life expectancy. Improving health care and its impact on death rates, birth rates and infantile mortality is clear. However, there are exceptions and they are worth looking at and explaining.

Table 1 Population data for selected countries, 2005 (1999 in brackets)

Country	Population (in millions)	Life expectancy (years)	Crude death rate (per 1000)	Crude birth rate (per 1000)	Children per woman	Infant mortality rate (per 1000 live births)
Afghanistan	29.9 (22.8)	42 (46)	22 (22)	48 (50)	6.8 (6.8)	17.2 (n/a)
Vietnam	83.3 (78.7)	72 (68)	6 (6)	19 (20)	2.2 (2.6)	18 (37)
India	1103.6 (998)	62 (63)	8 (9)	25 (26)	3 (3.1)	60 (71)
China	1303.7 (1253.6)	72 (70)	6 (7)	12 (16)	1.6 (1.8)	27 (30)
Bangladesh	144.2 (127.7)	61 (58)	8 (9)	27 (28)	3 (3.1)	65 (61)
Chad	9.7 (7.5)	47 (47)	17 (16)	45 (45)	6.3 (6)	101 (101)
Sierra Leone	5.5 (4.9)	40 (38)	24 (25)	47 (45)	6.5 (6)	165 (168)
Burkina Faso	13.9 (11.6)	44 (71)	19 (19)	44 (44)	6.2 (1.2)	81 (105)
Burundi	7.8 (6.7)	49 (43)	15 (20)	43 (41)	6.8 (6.8)	67 (105)
Rwanda	8.7 (8.3)	44 (41)	18 (22)	41 (45)	5.7 (6.1)	107 (123)
Egypt	74 (67.2)	70 (67)	6 (7)	26 (26)	3.2 (3.3)	37 (47)
Uganda	26.9 (21.5)	48 (40)	15 (19)	47 (46)	6.9 (7.1)	88 (880)
Malaysia	26.1 (22.7)	73 (72)	5 (4)	26 (24)	3.3 (3.1)	10 (8)
Singapore	4.3 (4)	79 (77)	4 (5)	10 (13)	1.3 (1.7)	1.9 (3)
Thailand	65 (60.8)	71 (69)	7 (7)	14 (17)	1.7 (1.7)	20 (28)
Mexico	107 (97.3)	75 (73)	5 (5)	23 (27)	2.6 (2.7)	25 (29)
Brazil	184.2 (167.9)	71 (67)	7 (7)	14 (17)	1.7 (1.7)	27 (32)
Russian Federation	143 (147.1)	66 (67)	16 (14)	11 (9)	1.4 (1.3)	12 (16)
Bulgaria	7.7 (8.2)	72 (71)	14 (14)	9 (8)	1.3 (1.2)	11.6 (14)
Germany	82.5 (82.1)	79 (77)	10 (10)	9 (9)	1.3 (1.3)	4.3 (5)
France	60.7 (58.8)	80 (78)	8 (9)	13 (13)	1.9 (1.7)	3.9 (5)
Italy	58.7 (57.6)	80 (78)	10 (10)	9 (9)	1.3 (1.2)	4.8 (5)
Japan	127.7 (126.6)	82 (80)	8 (8)	9 (10)	1.3 (1.4)	2.8 (4)
USA	296.5 (278.2)	78 (77)	8 (9)	14 (15)	2 (2)	6.6 (7)
UK	60.1 (59.8)	78 (77)	10 (11)	12 (12)	1.7 (1.7)	5.2 (6)

(Souce: World Population Data Sheet 1995 and 2005)

You should be able to draw quick sketches of the following global demographic characteristics from your own textbooks, research and learning:

- net population change
- crude birth rate
- infant mortality rate
- death rate
- life expectancy

More important than the actual distributions and figures are the reasons for them. Figure 5 illustrates the key factors explaining global variations in demographic indicators. The main factor is the **stage of development.** This in turn is the product of the factors shown in the diagram. Remember, the indicators of demographic health and all of the data are fixed in time. The reality is that countries and regions are all dynamic and changing. Countries develop both economically and socially, and this affects population change over time.

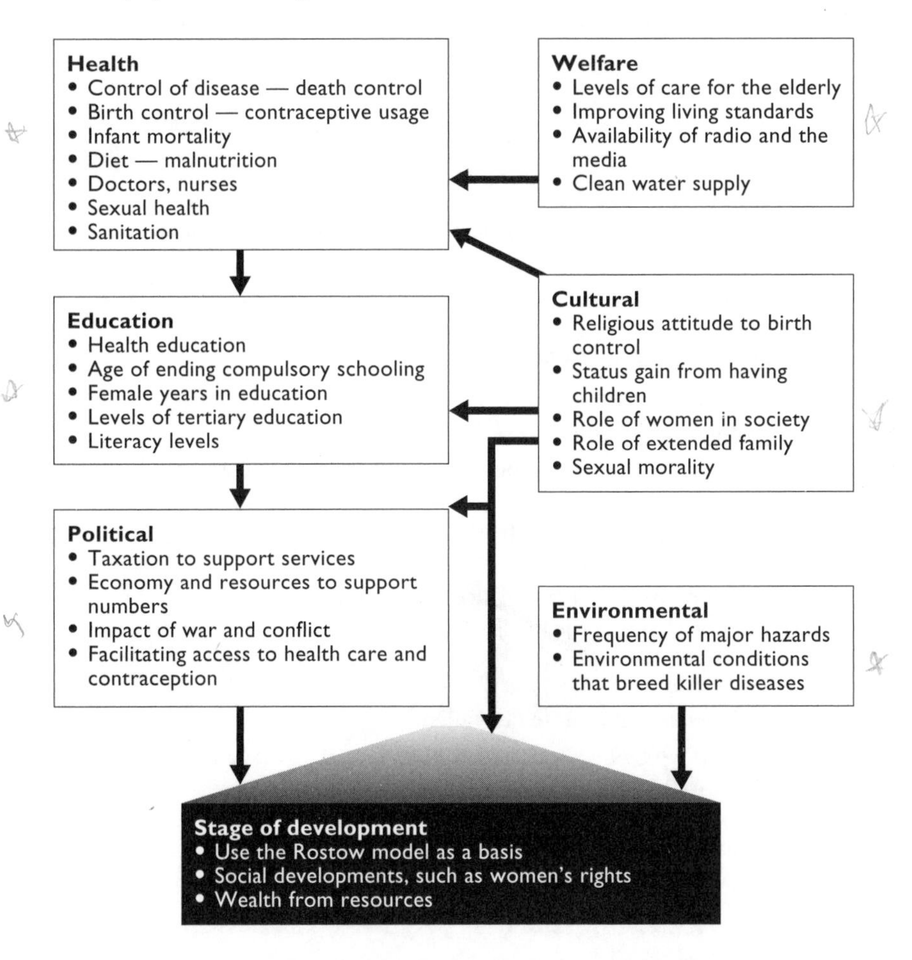

Figure 5 Explaining global variations in demographic indicators

1.2 How and why do populations change naturally?

The most common explanation for change is the **demographic transition**, a process of five stages (see Figure 6). Countries have distinctive characteristics at each stage. No country is now in Stage 1 although there may be peoples within some countries who are still at Stage 1. A key point is that the model ignores migration which is significant for many countries.

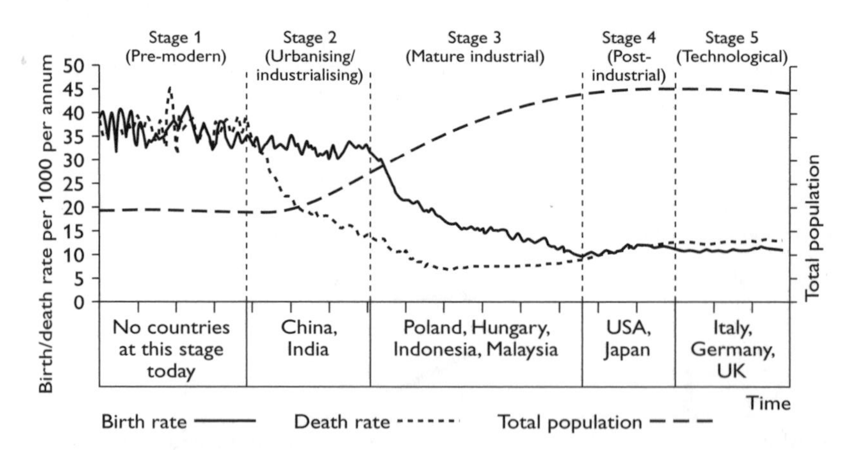

Figure 6 Five-stage demographic transition model

Stage 2 (e.g. Cape Verde; see also Figure 7)

- Population 524,000 in 2008.
- Accelerating growth with a large gap between birth rate (23.8/000) and death rate (6/000) in 2009.
- Medical revolution enabling death control.
- Lowering infant mortality (41/000).
- Increasing urbanisation and industrialisation — 60% urban.
- Out-migration — more Cape Verdeans live outside than inside the country.

Stage 3 (e.g. Chile; see Figure 8)

- Population growth because birth rate has fallen.
- Social customs have changed.
- Greater materialism and altruism — concern for life chances.
- Economic changes have encouraged fewer children to be born.
- Economy diversifies into services.
- Chile probably approaching Stage 4.

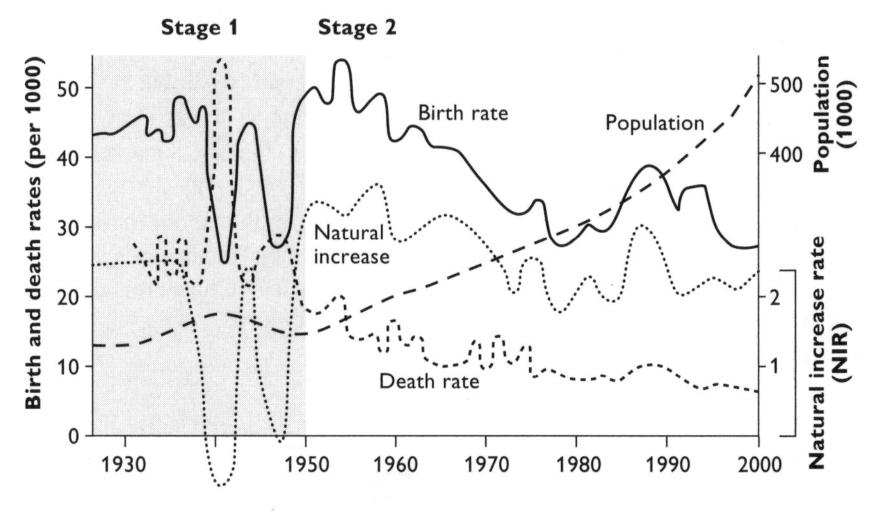

Figure 7 Cape Verde birth and death rates, 1930–2000

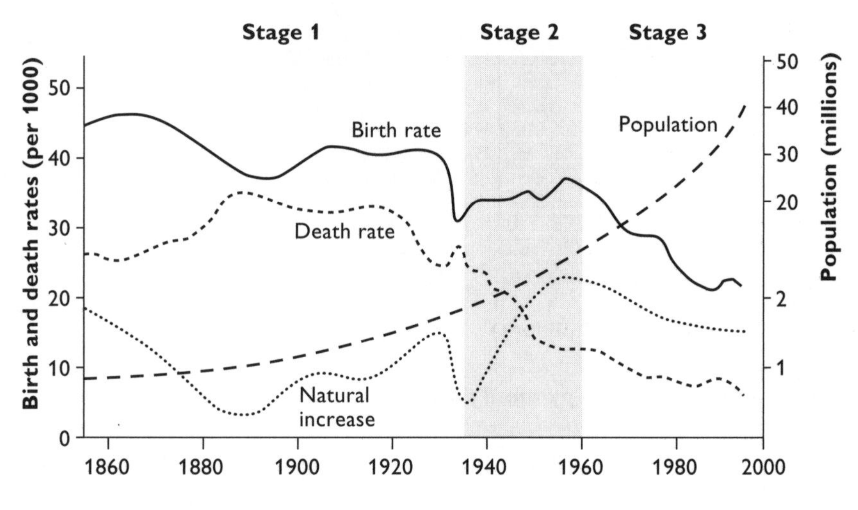

Figure 8 Chile birth and death rates, 1860–2000

Stage 4 (e.g. Denmark)

- Low, if not zero, growth.
- More women in the labour force.
- Lifestyle changes — higher incomes, more leisure and increased birth control.
- Cohabitation and delayed parenthood.
- In-migration keeps numbers from total decline.
- Post-industrial society entering Stage 5.

Stage 5 — the second demographic transition (e.g. Germany; see Figure 9)

- Noted since 1990s by European geographers.
- Very low birth rate and low numbers of children per woman (see Table 1).
- MEDCs with increasingly old-age populations, immigration to fill gaps in workforce and higher emigration of young people and highly skilled to the global marketplace.
- Economy is information, ICT and biotechnology based.
- Rise of individualism linked to emancipation of women in the labour market, new attitudes to contraception and abortion, and greater financial independence.
- Greater environmental concerns for the impact of increased numbers on resources.
- Rise of non-traditional lifestyle, such as same-sex relationships.
- Rise of childlessness and delayed pregnancy.

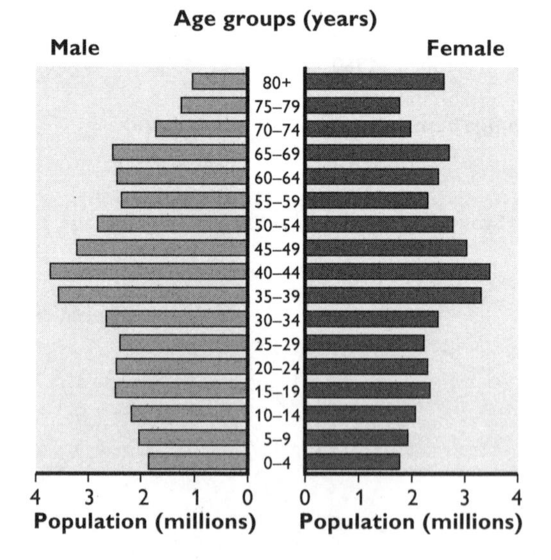

Figure 9 Germany: population pyramid for 2005

You can find evolutionary population pyramids for almost all countries at **www.nationmaster.com**

Implications of population change

World population is growing **exponentially** or **geometrically** — the number increases at a constant proportion of the number at a previous time. It appears as an increasingly steep curve on normal graph paper and a straight line on logarithmic graph paper. Table 2 tabulates our concerns by continent, using forecast data for 2050.

To cope with population growth there have to be resources at the global, continental or national level to support the population. The ability to support a population is measured by the **population–resource ratio**, a dated attempt to link the ability of a population to support itself from its own resources and by trading.

Table 2 Population growth, 1950–2050

	Population (millions)			Share of world population (%)			Change (millions)		Share of world change (%)	
Continent	**1950**	**2005**	**2050**	**1950**	**2005**	**2050**	**1950–2005**	**2005–50**	**1950–2005**	**2005–50**
Africa	221	906	1,969	8.8	14.0	21.2	685	1,063	17.2	38.4
Asia	1,402	3,921	5,325	55.6	60.4	57.5	2,519	1,404	63.3	50.7
Europe	547	750	660	21.7	11.5	7.1	203	–90	5.1	–3.2
Latin America	167	559	805	6.6	8.6	8.7	392	246	9.8	8.9
North America	172	329	457	6.8	5.0	5.0	157	128	3.9	4.6
Oceania	13	33	46	0.5	0.5	0.5	20	13	0.5	0.4
World	**2,522**	**6,498**	**9,262**	**100**	**100**	**100**	**3,976**	**2,764**	**100**	**100**

Sustainable population, once called **optimum population,** is the population that can be supported so that natural resources are not depleted and output is maximised within the prevailing technological, economic and social conditions. It is a theoretical aim for the world and more so for individual countries.

The major concern for the future is **overpopulation**. This is where resources are unable to sustain a population at the existing living standard without a reduction in that population or an increase in the resources available to support it. It is a concern for LEDCs where Malthusian checks such as famine, AIDS and war have an impact.

Underpopulation is rarely a concern today, although it is becoming one in some MEDCs where populations are declining naturally. International migration is often encouraged in these circumstances. It occurs when existing resources could sustain a larger population without lowering living standards, or where the population is too small to develop the available resources.

The debate over the ability of the world to support its population has been around since Thomas Malthus wrote his theory in 1798. In the twentieth century, Esther Boserup (1965) and the Club of Rome (1972) respectively challenged and supported Malthus (see Figure 10). **Neo-Malthusians** have suggested that famines in the Sahel have shown that population is outstripping food supply. Even the Food and Agriculture Organisation (FAO) suggests that up to 800 million people are malnourished. Population growth is accelerating in Africa and impeding growth although the AIDS pandemic has slightly slowed the increases. Water shortages partly caused by climate change are predicted to become a major issue by 2050. Today, we are more aware than ever of the finite nature of the world's resources. This has consequently increased the importance of **sustainable development** (introduced to the world at the 1972 Global Environmental Summit in Stockholm), and the critical links between population, the economy and the environment. The 2002 World Summit in Johannesburg focused upon sustainable management of the global resource base, poverty eradication and improved healthcare.

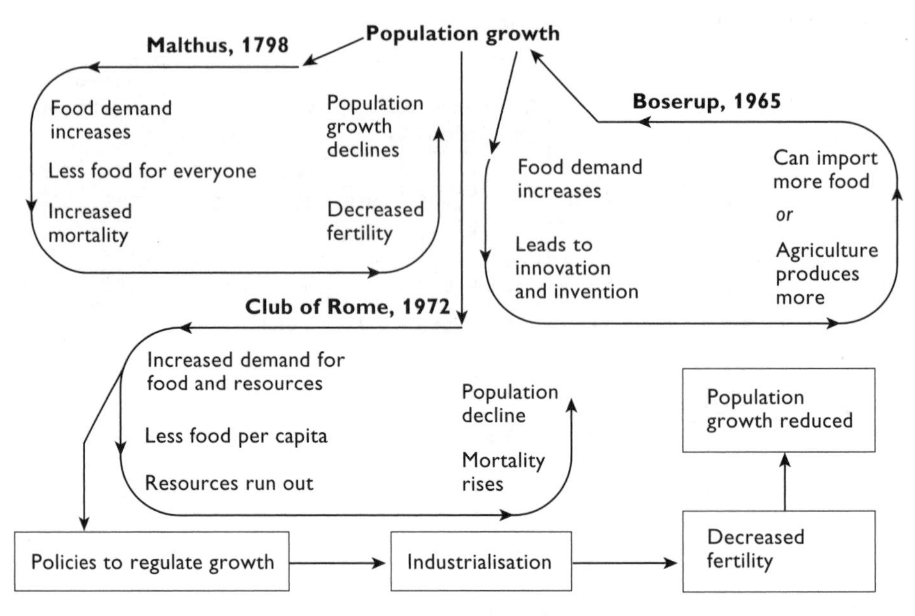

Figure 10 Three models of the impact of population growth

1.3 What is the role of migration in population change?

Motives — why do people migrate?

Figure 11 provides a basic model of migration at any scale. The balance between immigration and emigration is called **net migration**. It will vary over time within a country and between countries.

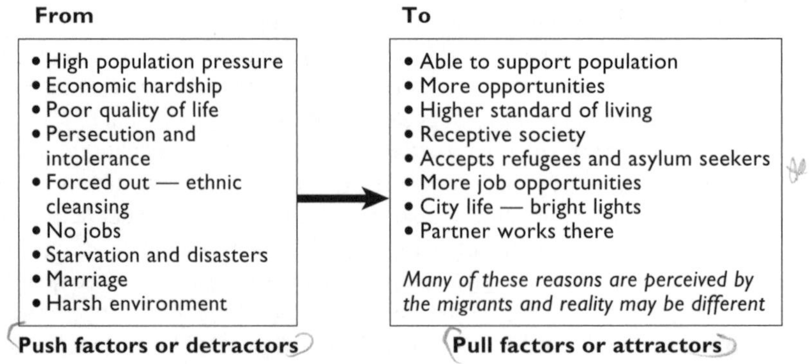

Figure 11 Motives for migration

Duration

Migration is generally long term and mostly for economic reasons. However, much publicity has been given to refugees and asylum seekers who are often, mistakenly, grouped with economic migrants.

	Short term						Medium term					Long term
	Movement					Migration						
Type	Leisure	Commuting	Weekend	Holiday	University	Job moves	House moves	Expatriot	Retirement	Refugee	Asylum	Emigration
Timescale	Part-day	Day	2 days	2 weeks	3 years	Several years				Temporary to life		Life
Typical	Normally within a country				Possibly international		International					

Figure 12 Duration of migration

Voluntary or forced migration?

Most migration is the result of free choices made by the migrant, the head of a migrant household, or a group. However, according to the UNHCR, 22 million people have been forced to leave their homes and/or country. See Table 3 for examples.

Table 3 A typology of international migration

Voluntary	Examples	Forced	Examples
Between MEDCs	Italians to Bedford in the 1950s: brickworks. Poles to work in UK after 2004.	Between MEDCs	Ethnic Germans from Hungary to West Germany in 1945
Skilled labour	Financial experts to New York and Singapore	n/a	n/a
Between LEDCs	Lesotho to South Africa	Between LEDCs	Hutus from Rwanda to DR Congo. Palestinians to West Bank and Gaza
LEDC to MEDC	Bangladeshis to UAE	n/a	n/a
Labour migrants	Caribbean to UK in 1950s. Mexicans to USA	Labour migrants	Sex industry workers from Eastern Europe
Refugees	Montserrat to Antigua following volcanic eruption	Refugees	Ethnic cleansing in former Yugoslavia. Tamils from Sri Lanka to India
Asylum seekers	Afghans to Australia/UK, Iraqis to UK	Asylum seekers	Kurds from Iraq to Italy

Models of migration

Ravenstein's laws

Ravenstein's laws are based on Britain in the 1880s. He gave migration the following characteristics:

- it is short distance
- it is step-by-step
- longer distances are travelled to major centres
- there are reverse flows
- rural people are more inclined to migrate than urban people
- females move more than males
- most migrants are adults
- large towns grow more by migration than by natural increase
- volume grows as industrialisation progresses
- the major flow is from agriculture to industrial towns
- the major cause is economic

Do these 'laws' apply today? If so where and why? If they do not, why is this so?

Lee's model

According to Lee people:

- assess and perceive the destination
- assess conditions where they are
- look at obstacles between the two, such as distance and cost
- consider personal circumstances

Stouffer's intervening opportunities model

This model (Figure 13) attempts to explain why migrants settle for locations other than their original, intended destination.

Figure 13 Stouffer's intervening opportunities model

Todaro's model

Todaro's model is based on two principles:

- Economic factors are the most important of the push–pull factors.
- Migration is most likely where urban incomes are greater than rural incomes.

The decision-making model

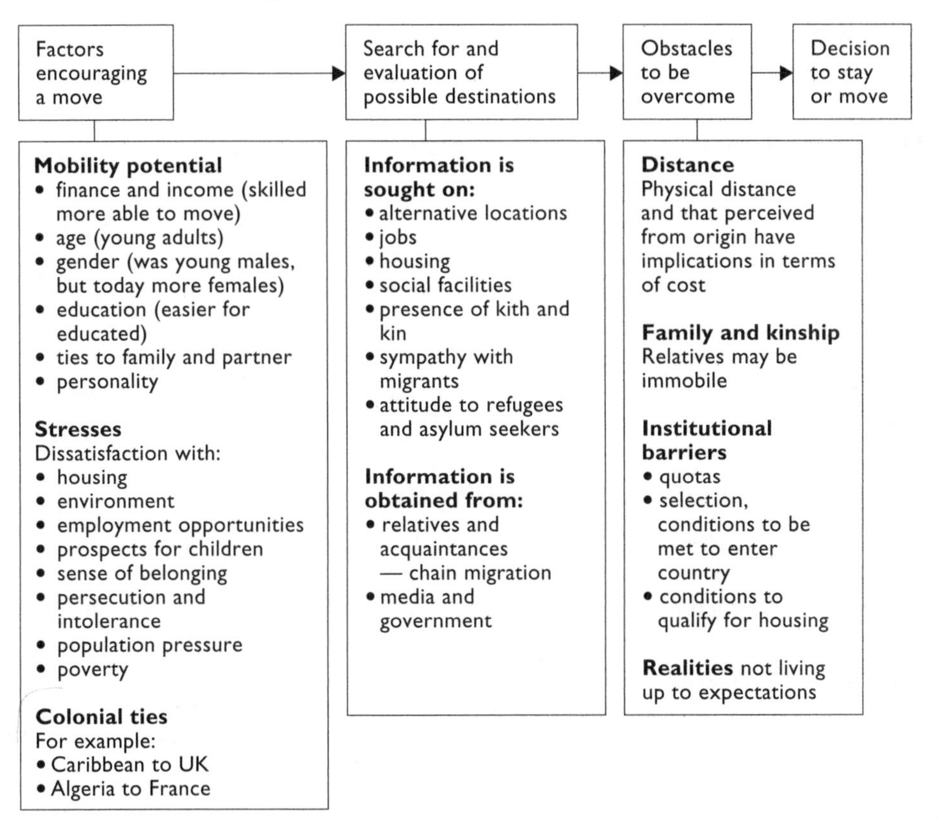

Figure 14 The decision-making model

Migration at a national scale: the UK

In the nineteenth century people migrated to the coalfields, for example in South Wales from rural Wales and Ireland. In the first half of the twentieth century there was movement towards London and the South East. Since the 1930s, the process of **suburbanisation** has seen movement to the outer edges of cities. **Counter-urbanisation** in the late twentieth century continues today, and is a net movement of people away from cities to small towns and villages. There are other forms of migration: retirement migration to southwest England, migration to live in the Dordogne or Spanish Costas following second-home purchase, and student migrations to university towns and then away again on graduation. There is significant **re-urbanisation** (migration back into cities) by mainly high-flying single people and the wealthy.

In 2001, 9% of the UK population was born outside the country: 5% Asian or British Asian, 2% black or black British, 1% Irish and 1% EU citizens. The issues in the UK are:

- Media portrayal of immigration using emotive language, for example, 'flood', 'tidal flow' and 'swamping' of services.
- Difficulty of distinguishing economic migrants from refugees and asylum seekers.

- Migrant communities encouraging chain migration of relatives to the same area.
- Issues of controlling the UK border.
- Pressure on housing, education, health service and social services.
- Much of the debate is clouded by the issues of refugees and asylum seekers.

The impact of international migration

More countries are being affected by international migration as the numbers migrating grow. There is more variety among migrants and no type dominates the flows as in the past. More women are labour migrants in their own right. These details pose new challenges both for receiving and exporting states. More migration is short term as states place limits on work permits. Long term migration has declined because of political pressures from the host population, the lack of low skilled jobs, and tightened entry requirements such as the UK points system — the **Highly Skilled Migrant Programme**. Although people migrate for work, many return home later.

Impact on the country accepting the migrants

- Demographic replacement for low natural change and declining population, for example Turks to West Germany in the 1970s, Croatians to Switzerland and Iranians to Sweden (although some sources may class these as refugees). In 2002, Germany recognised the need for 50,000 migrants per year as a demographic replacement for retiring Germans.
- Labour force needs are satisfied, e.g. Caribbeans to the UK in the 1950s, Portuguese (Madeirans) to Guernsey in the 1990s and more recently Latvians. Some of these movements become permanent because of marriage but many comprise **guest workers** — people recruited for a period of time such as South Asians to the UAE, or employees on assignment such as workers from Barclays Bank and Microsoft in the UK working in New York and Seattle.
- Increased pressure of population on resources, such as food and land, housing and social resources. This is the argument made by right-wing groups in the UK.
- It helps to create a **multicultural society**, e.g. Toronto. The religious mix, retailing, restaurants and music of the host country may alter.
- Areas may become dominated by an immigrant group, e.g. Turks in Cologne, Poles in Illinois, Caribbeans in New York and New Jersey, and the Chinese in the west coast cities of the USA creating vibrant Chinatowns.
- There are issues of race relations and **segregation** such as the divide in Blackburn between immigrant peoples on one side of town and the long-standing residents on the other. **White flight** and **ghettoisation** may be further outcomes.
- Political reactions such as BNP, National Front in France, Austrian Freedom Party, Vlaams Blok in Belgium.
- Gender concentrations, e.g. south Asian men in the UAE and Mexican women in the USA.
- Illegal labour on low wages, e.g. Mexicans on Californian farms, and some of the gangmaster cases in the UK.

Impact on the country sending the migrants

- Slowing down of natural increase because the fertile migrate.
- Old-age society, as young adults leave.
- Slightly decreased pressure on resources, e.g. after the Irish potato famine.
- Fewer people to engage in agriculture and produce food.
- Receipt of **remittances** from workers to family back home. However, the flow of remittances can decline, e.g. from USA to Mexico in 2009, resulting in hardship for the family back home. Remittances fell by 18% in 2009.
- **Return migration** with new skills and prospects, e.g. taxi drivers and chefs to the Mediterranean resorts.
- **Westernisation** and **cultural imperialism** reinforced by returnees.
- Solution to a political or racial issue, e.g. ethnic cleansing in the Balkans, and Ugandan Asians in the 1970s who left for the UK.
- Loss of skills, e.g. IT experts moving to USA from India.
- Pensions outflow where old-age migration is taking place, e.g. UK pensions going to Spain and Dordogne.
- Return migration during hard times in the receiving countries.

Return migration

Economic downturns such as the 2008–09 credit crunch have the effect of encouraging many economic migrants to return to their country of origin. In 2009 it was noted that professional Nigerians and Ghanaians working in the UK were returning home, a phenomenon called **brain gain.** Similarly, there is a return flow from the USA to Mexico but of a mixture of skilled and unskilled workers. The effect is a reduction of income for Mexican families. For other labour migrants, returning home was part of their original plans once they had made money. Throughout the past 40 years there has been a steady return migration from Germany and Switzerland to the countries of southern Europe. Many returnees had acquired the capital to set themselves up in business in their home area.

Migrant labour

In 2001, Jersey had 38,662 persons born outside the island. British born form 36% of the population and are declining in numbers. Portuguese and especially Madeiran born were 6.4% of the population (5,135 persons); Irish born were 2.6% (2,840). The population pyramid for the Portuguese is typical of labour migrants, with 54.6% being aged between 20 and 39 years old. Males slightly outnumber females in these age groups. There are slightly more male migrants, although females dominate the Irish migrants especially in the 20–29 age group.

These differences probably reflect a number of factors:

- The Irish have a language affinity with the islanders and were earlier migrants.
- Distance from the home country has meant that males were more likely to be the dominant Portuguese migrants. In addition, many Portuguese migrants come from islands such as Madeira and the Azores that are even further away.

- It is mainly a singles population judging from the small number of dependents and retired migrants.
- Both Ireland and Portugal have higher unemployment and lower living standards than Jersey.

The Portuguese are more recent short-term migrants, often returning home at the end of the tourist season. The same characteristic is repeated but less strongly, for the Irish. Both groups are employed in lower skilled work, particularly in hotels, although some of those who have settled or keep returning to the island have risen to the status of managers in the holiday industry. UK immigrants are a mixture of affluent people seeking a tax haven and skilled workers attracted to work in the finance sector and government employment, such as education. Therefore, the UK born are less likely to be recent migrants. Among the consequences of the Portuguese presence are that:

- shops in St Helier have Portuguese speaking assistants to serve the immigrants
- the BBC in Jersey has Portuguese language broadcasts
- the Catholic Church has services in Portuguese

Other cases where guest workers make an important contribution to the national economy are: Switzerland (26.5% of the workforce in 2007), Germany (7.5% of the workforce), the Gulf States such as UAE, Bahrain and Kuwait where many males have migrated from India, Pakistan and Bangladesh. Similar movements can be found from Indonesia to Malaysia and from the Philippines to Singapore, Malaysia and Hong Kong.

Table 4 summarises some of the costs and benefits of migrant labour to both the source and destination countries.

Table 4 Issues resulting from labour migration

Source country		Destination country	
Economic costs	**Economic benefits**	**Economic costs**	**Economic benefits**
Loss of young adult labour to source country	Less underemployment	Costs of educating children	Many less desirable jobs taken in host country
Loss of skilled labour may slow development	Return migrants (reverse or counter-migration) bring skills back home	Some skills may be bought cheaper, so home people are disadvantaged	The host country may gain skilled labour at little cost; can fill skills gap (e.g. programmers from India to USA)
Regions of strongest out-movement suffer from spiral of decline (e.g. parts of inland southern Italy)	Funds repatriated to source country	Funds repatriated to source country; pension outflows following retirement to home country	Some retirement costs transferred back to source country
Loss of skilled labour may deter inward investment	Less pressure on resources such as food	Slightly greater pressure on resources	Dependence of some industries on migrants (e.g. construction)

Source country		Destination country	
Social costs	**Social benefits**	**Social costs**	**Social benefits**
Encourages more to migrate, with effect on social structure (e.g. Ireland in the 1840s)	Reduced population density; lowers birth rate because young tend to migrate	Discrimination against ethnic minorities (e.g. riots in Bradford, 2001); political unrest and opposition from parties of the right (e.g. in Austria)	Creation of multi-ethnic, multicultural society – building cultural facilities
Dominance of females left behind	Funds sent home can provide for improved education and welfare	Dominance of males, especially in states where status of women is low (e.g. Gulf States)	Greater awareness of other cultures (e.g. Caribbean music, literature and religion in the UK)
Non-return can unbalance population pyramid	Retirees building homes	Loss of aspects of cultural identity, especially among second generation	Providers of local services (e.g. Turkish baths and UK newsagents)
Returning on retirement is a potential cost	Some able to help develop new activities (e.g. tourism)	Creation of ghettos (e.g. Turks in Berlin); schools dominated by immigrants	Growth of ethnic retailing and restaurants (e.g. Rusholme, Manchester)

1.4 Refugees and asylum seekers

It is difficult to distinguish between the following:

- **Economic migrants** — people seeking opportunities and legitimately entering a country.
- **Refugees** — those genuinely fleeing from persecution and harsh political regimes. A refugee is defined by the United Nations (UN) as a person unable or unwilling to return to his/her homeland for fear of persecution based on reasons of race, religion, ethnicity, membership of a particular social group, or political opinion.
- **Illegal immigrants** — people who enter a country without authority and hope to remain.
- **Asylum seekers** — people who request refugee status in another state.

Cases such as Cubans fleeing to the USA during the past 40 years and the Vietnamese boat people in 1975 point to the dilemma, as do Afghans today trying to board lorries passing through the Channel Tunnel.

The UN projection is that by 2050, the MEDC population will not change much whereas there will be an extra 3.1 billion people who inhabit LEDCs. The population of Europe will decline by 5% by 2025 and therefore the labour force will have to be

replaced either by a higher rate of natural increase or by the immigration of economic migrants, refugees and asylum seekers. There have been refugees throughout history. Examples includes:

- Persons displaced in post-1945 Europe — for example, Germans in Moldova.
- Movements between India and Pakistan in 1947 upon independence of the two countries.
- Vietnam in 1975 when 800,000 boat people fled — some are still in Hong Kong.
- Myanmar 1991–92: 250,000 Muslims fled and 21,000 remained in Bangladesh.
- Sri Lanka 1992–present: Tamils fleeing to India where at least 64,000 still live.

In 2009 Kosovo, Pakistan, Somalia, Sudan, Chad and DR Congo all have problems giving rise to refugee movements.

Refugees

The UNHCR estimated that there were 50 million people displaced forcibly from either their home areas (**internally displaced persons** (**IDPs**)) or their home countries. In 2009 there were 10.5 million refugees. The largest group were 4.7 million refugees in camps in the Middle East. UNHCR's African refugee emergency areas are concentrated in Central Africa: Chad, Central African Republic, Sudan, Uganda, DR Congo, Tanzania, Ethiopia and Somalia.

Some countries (e.g. Norway, Sweden, Finland and Switzerland) have a long tradition of accepting refugees and asylum seekers who are unable to stay in their intended country of refuge. Canada adds gender persecution to its criteria for accepting refugees. Figure 15 shows the major source countries of refugees in 2007.

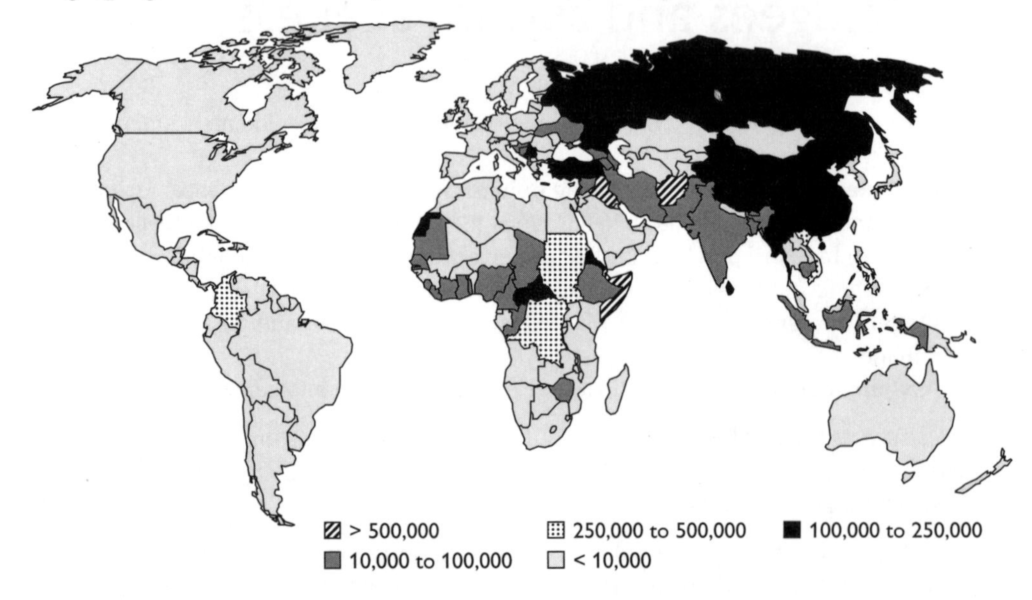

***Figure 15* Major source countries of refugees, end 2007**

Rwanda and Burundi

In 1994 in Rwanda, the **genocide** of at least 500,000 people displaced forcibly as a consequence of civil war between the Tutsi and Hutu peoples resulted in the movement of 1.75 million refugee Hutus to neighbouring countries. As Hutus left, a counter flow of 700,000 Tutsi refugees began. The UNHCR encouraged repatriation and about 160,000 Hutus returned. But the high birth rate among the refugees actually meant that the population remaining outside Rwanda stayed the same. The refugees were housed in vast camps near the Rwandan border in DR Congo and Tanzania. In 2003 the refugees in Goma were forced to move again by the eruption of Mount Nyiragongo.

In Burundi, a similar tribal conflict, commencing in 1993, was made worse by the flight of the Hutus from Rwanda. Many were forced back or fled to Tanzania. The political situation in eastern Congo, where civil war was raging, resulted in the further displacement of 400,000 Rwandans and Burundians into DR Congo. The Congolese forced 650,000 to return to Rwanda and Burundi. By 1996, 1.7 million Rwandans and 200,000 Burundians were in camps. At the end of the crisis in 1997, 600,000 were repatriated from Tanzania and the Congo but 215,000 remained unaccounted for.

The crisis resulted not only in the deaths of over a million people but also in environmental degradation, especially around the sites of the refugee camps, which were denuded of timber. War and flight had halted all development. Even in 2009 many still live in poverty in camps. In 2007, an estimated 300,000 refugees and 1.7 million IDPs made the decision to return home, mainly to Burundi and DR Congo.

Refugees in the UK

In 2009, only 20% of the estimated 15.2 million refugees reached developed countries. At the end of 2008, the UK hosted 292,100 refugees whereas Germany hosted 582,700. The UK is not the most receptive country and is only tenth in Europe for the number of refugees as a percentage of its population. Therefore, you have to check the accuracy of media statements on refugees. The majority (some estimate 1 in 20) live in London. Newham, Enfield, Haringey, Redbridge and Waltham Forest are the boroughs with the highest totals (including asylum seekers). All but Haringey are in east London — traditionally the area of the capital that receives migrants.

Illegal immigrants

There are an estimated 2.6 million illegal immigrants in Western Europe. Nearly all countries are attempting to tighten immigration controls (e.g. UK Border Agency), often in the wake of xenophobic (right-wing nationalist) attacks on refugees and 'illegals'. In 2009 there were estimated to be 725,000 illegal immigrants in the UK although the very nature of illegal immigration means that the figure could range between 524,000 and 947,000. The destinations of illegal immigrants are not easy to ascertain. They spread widely and are probably distributed where the illegal employment opportunities are best — in low-grade agricultural work such as vegetable picking, the hotel industry as cleaners, and general labouring jobs. London is disproportionately affected. In 2007, the estimates showed that there were 518,000 illegal immigrants in London. Only 111,265 illegal immigrants have been deported in 10 years since 1998. Deportation represents a cost for the country.

Asylum

Asylum may be defined as 'the formal application by a refugee to reside in a country when they arrive in that country'. The number of asylum seekers is invariably larger than the number who request refugee status. The numbers seeking asylum have increased steadily since 1970, just when many countries decided to curtail immigration. One catalyst was the German realisation that Turkish guest workers were unlikely to return home. The collapse of the former Yugoslavia in 1992 and the subsequent civil unrest in the region also resulted in a rise in asylum seekers from Bosnia-Herzegovina, Croatia, Serbia and Montenegro. Figure 16 shows the origins of global asylum seekers in 2008, which totalled over half a million worldwide.

There were 25,000 applicants for asylum in the UK in 2008, a 10% rise on 2007. Of these, 19% were granted asylum, 11% granted either humanitarian protection or discretionary leave and 70% were refused. In the same year, 66,275 persons were removed or departed voluntarily from the UK. Table 5 places the UK in its European context.

Figure 16 Country of origin of new asylum seekers in 2008

Table 5 Asylum applications in Europe, 2008

Country	Number	Applicants per million inhabitants	Percentage of European (EU) total
France	41,845	180	16.3
Italy	30,145	80	11.7
Germany	26,945	180	10.5
UK	25,000	no data	9.7
Sweden	24,845	580	9.6
Greece	19,885	420	7.7
Switzerland (not EU)	16,505	650	
Belgium	15,940	no data	6.2
Netherlands	15,255	no data	5.9
Norway (not EU)	14,430	810	
EU 27	256,405	130	100

The flow of asylum seekers is to the industrialised countries. The EU destinations in Table 5 receive 78% of all EU asylum applications. The pattern can be explained as follows:

- Proximity — Italy for Albanian and African applicants, Greece for Balkan, Turkish and Kurdish applicants.
- Receptiveness — Scandinavian reputation for compassion and receiving migrants.
- Neutrality — Switzerland and Sweden.
- Perceived job opportunities — Germany, France, UK.
- Links as a result of colonialism — Zimbabweans to UK, West Africans to France.

Reasons for increased prominence of asylum seeking

- Pressures to migrate from the poorest states are increasing due to a combination of economic decline and political instability.
- Improved communications enable people to learn more about destinations and the means to reach them.
- Communities already exist in destinations that have encouraged others to join them. (Chain migration involves members of an extended family following one another.)
- The costs of transport have declined.
- More gangs of human traffickers prey on would-be migrants and offer passage to a new life.
- Destination countries find it difficult to distinguish between those fleeing from threats to their life and liberty, and those trying to escape poverty and improve their quality of life.

Costs of asylum seeking in destination countries

Increased costs associated with asylum seeking include:

- housing
- social services
- schooling
- welfare payments
- 'improved' border controls at points of entry
- dispersal costs to areas where housing exists
- policing of demonstrations against these people by right-wing extremists

These costs have to be paid for from taxation. Asylum seekers often wish to live in the heart of a country such as the capital city and do not want to go to depressed areas while their application is processed.

Policy reactions

(1) Limiting seekers at source with more difficult visa requirements.

(2) Pre-boarding arrangements to prevent departure unless the passenger has a return ticket.

(3) Prevention of illegal crossings by physical constraints, e.g. Channel Tunnel.

(4) Fast-track procedures for the genuine cases.

(5) Accepting those with skills, often using a points system.

(6) Maintain third country policies as used by Germany, whereby people entering the EU in East Europe intending to go to Germany were subject to German regulation in e.g. Bulgaria, the Czech Republic and Poland. If they had no valid German visa they were sent back from their point of entry.

(7) Temporary protection followed by repatriation once crisis subsides, which was used for Bosnians in 1996–97.

(8) Policies of NGOs aiding governments in the countries of both origin and destination to address the deep-seated causes of poverty.

Stateless people and ethnic cleansing

The UN Universal Declaration on Human Rights states that everyone has the right to a nationality. However, there are peoples for whom this right is disputed and who are denied the right to have a home and to work. The largest European group is the Roma, Gypsy population: there are 12 million in Europe (Romania 1.8 million and Bulgaria 700,000). In some cities they live in ghettos and their life expectancy, on average, is 15 years less than that in the host nation.

The break-up of Yugoslavia and the USSR resulted in some groups being declared stateless for political and nationalistic reasons (e.g. Russians in Latvia and Serbs in Croatia and Macedonia). The result has been **ethnic cleansing** — the forced and voluntary movement of peoples to states that are culturally compatible (having the same religion and/or language). Where the patchwork of groups has been too complex, there has been some internal ethnic cleansing to produce areas dominated by one culture.

Other cases of statelessness include 3 million Palestinians in Gaza and the West Bank, Vietnamese in Cambodia and Myanmari Muslims.

1.5 Changing gender structures

Gender is a generic term that refers to both males and females. Therefore, in any question related to gender you may refer to males as well as females. Women's role in society has demographic, economic, environmental and social causes and effects, including attitudes to work and child rearing, attitudes to female children, and attitudes to migration. There have been imbalances in gender roles because of historic causes. Traditionally in north east England there has been less work for women because the main employment was in coal mining and shipbuilding. In contrast, Lancashire had a tradition of female labour in the textile industry. These differences are less marked in the twenty-first century.

Gender and the demographic transition

In countries at Stage 2 of the demographic transition, women are regarded as mothers for children due to the high infant and childhood mortality rates. They are also the cooks and in much of sub-Saharan Africa women are the providers of agricultural labour. In Guinea Bissau women still collect 90% of the water. Depending on the country, sub-Saharan African women provide 60–80% of the agricultural labour. Others work in the informal, unregulated economy where wages are low. Women are more likely to be illiterate because they have not had the opportunity to be educated.

Some Stage 3 countries use women as cheap labour. In Asia up to 50% of women provide agricultural labour and in some newly industrialised countries (NICs), such as Thailand and Indonesia, they are providing labour in food processing and other industries. The demographic transition has indirectly enabled more women to be educated.

In LEDCs women may lack self-esteem, employment other than in the home, schooling and the full and legal rights of citizenship. Some depend on children for status and security. Policies to improve women's rights include:

- Giving girls a good education, which attempts to empower women outside the home (Figure 17). Many girls, particularly in Africa, do not have equal access to schooling and so there is a particular need to provide secondary education in those countries.
- Providing employment or small business opportunities.
- Reducing the impact of patriarchal societies where children are regarded as a statement of virility.
- Reproductive health clinics to assist women to value their lives.
- Emphasis on contraception for both men and women.
- Getting tribal leaders to promote lower birth rates.

UNIFEM, the United Nations Development Fund for Women, is an organisation for women which was founded in 1976 and is attempting to foster empowerment through

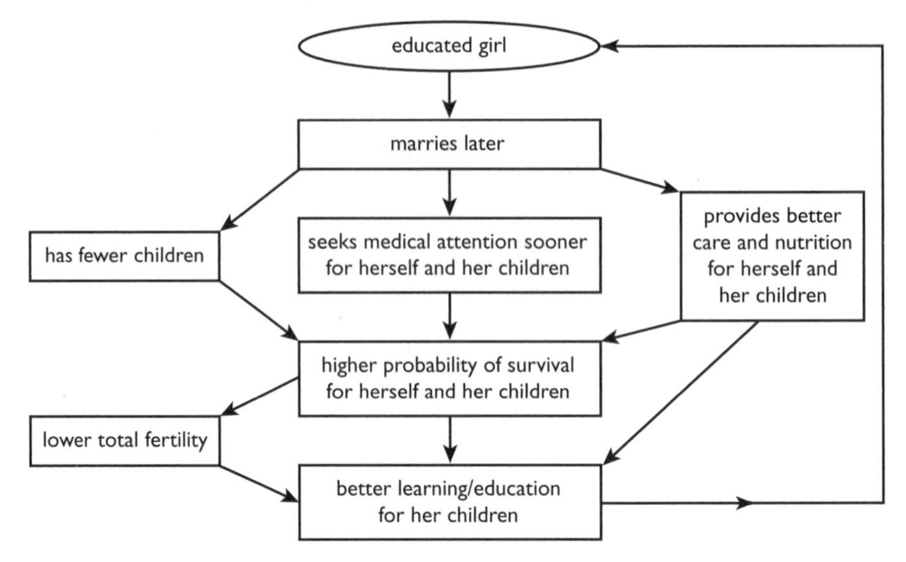

Figure 17 Possible demographic impacts of educating girls

gender equality. It has supported gender sensitive training in Jordan. It is also focusing on the role of women in the spread of AIDS in India. Currently, among 15–24-year-olds who have the AIDS virus worldwide, 50% are women. Counteracting the AIDS pandemic is a gender issue because it affects both men and women.

The Millennium Development Goals' Target 1B addresses women's issues where it aims to have full and productive employment and decent work for ALL. In addressing the issue of hunger and the 50% of people living on under US$1 a day it is addressing the needs of women in the eyes of UNIFEM.

The transition through Stage 3 still reflects a world dominated by men. Industrialisation and industrial urbanisation results in employment dominated by men. Transport developments such as railways and trams have been seen as providing opportunities for men to travel to work from the suburbs. Nevertheless, as we saw earlier, the textile industry did provide opportunities for women. The World Wars were a catalyst for changing attitudes to women at work. Suddenly, women were needed to keep factories operating (not least munitions factories), to work in the service sector (e.g. bus conductresses) and to work in agriculture (the Land Army in World War II).

The USA represents a good example of a country at Stage 4 of the demographic transition where, in 2008, 68 million women made up 46.5% of the labour force. They worked in the following areas:

- 39% in management and professional jobs
- 33% in sales and office work
- 21% in other service sector jobs
- 6% in industry and transport
- 1% in resource related jobs

Female employment did show racial variations: whites and Asians held 41% and 46% respectively of female management and professional positions whereas Afro-Caribbeans and Hispanics held 32% and 33% of sales and office jobs. The types of occupation still reflect a degree of gender bias. The 10 dominant occupations were:

(1) Secretaries and administrative assistants (3.2 million)
(2) Registered nurses (2.5 million)
(3) Elementary and middle school teachers (2.4 million)
(4) Cashiers (2.3 million)
(5) Retail sales (1.8 million)
(6) Psychiatric nursing and home health aids (1.7 million)
(7) First line supervisors and managers (1.5 million)
(8) Waitresses (1.4 million)
(9) Receptionists and information clerks (1.3 million)
(10) Book keeping, accounting and auditing clerks (1.3 million)

Women outnumber men in the USA in, for example, public relations, human resources, education administration, health management, teaching and nursing. This enhanced role is partly due to the educational opportunities that are available. Almost a third of both genders have degree level qualifications.

Stage 5 is difficult to relate to gender development because those countries with declining populations, such as Italy, are not ranked very highly on the GDI (Table 6). However, Stage 5 countries have a low birth rate, and a high degree of female emancipation and contraceptive use, which when placed alongside delayed pregnancy, small family size and same-gender unions, is a distinctive aspect of the stage. In Stages 4 and 5 women live longer and outnumber men in the older age groups. War losses can affect the balance between males and females as it did markedly in Germany after the two World Wars. Many still see **glass ceilings** that prevent women, even in the most gender-equal societies, from attaining equality of opportunity for the top jobs.

The UNDP has attempted to measure progress on the **Gender-Related Development Index** (Table 6). The table does not recognise the demographic transition preferring instead to use a threefold classification of human development (high, medium, low) based on the **Human Development Index (HDI)** which you will come across if you study Development in Unit G3. GDI uses the dimensions of: (i) a long and healthy life indicated by life expectancy of men and women; (ii) knowledge indicated by adult literacy rates; and (iii) a decent standard of living indicated by estimated earned income. The higher the ranking, the greater is gender equality: 1 is total equality. Use the table to provide your own examples of gender differences. Ask yourself whether these figures reflect the demographic transition.

Gender and migration

Migration affects gender balance in the areas of origin and destination. For every 2.5 Mexican males entering the USA, only 1 woman makes the same journey. What is the effect on the areas of origin? The African boat people making the hazardous crossing from Libya to Malta and Lampedusa (Italy) are predominantly male.

Table 6 Selected country data on Gender-Related Development Index (GDI), 2005

HDI rank	GDI rank and value	Life expectancy at birth in years		Adult literacy rate (15 and over)		Estimated earned income (1000 US$)	
High human development		F	M	F	M	F	M
1 Iceland	1 0.962	83.1	79.9	100	100	26.6	40.0
2 Norway	3 0.957	82.2	77.3	100	100	30.7	40.0
7 Switzerland	9 0.946	83.7	78.5	100	100	25.0	40.0
12 USA	16 0.937	80.4	75.2	100	100	25.0	40.0
16 UK	10 0.944	81.2	76.7	100	100	26.2	40.0
20 Italy	17 0.936	83.2	77.2	98	98.8	18.5	39.1
61 Saudi Arabia	70 0.783	74.6	70.3	76.3	87.5	4.0	25.7
Medium human development							
73 Kazakhstan	65 0.792	71.5	60.5	99.3	99.8	6.1	9.7
74 Venezuela	74 0.787	76.3	70.4	92.7	93.3	4.6	8.7
81 China	73 0.776	74.3	71.0	86.5	95.1	5.2	8.2
121 South Africa	107 0.667	52.0	49.5	80.9	84.1	6.9	15.4
128 India	113 0.600	65.3	62.3	47.8	73.4	1.6	5.1
151 Zimbabwe	130 0.505	40.2	41.4	86.2	92.7	1.5	2.5
154 Uganda	132 0.501	50.2	49.1	57.7	76.8	1.2	1.7
Low human development							
158 Nigeria	139 0.456	47.1	46.0	60.1	78.2	0.6	1.5
167 Burundi	147 0.497	49.8	47.1	52.2	67.3	0.6	0.8
168 DR Congo	148 0.398	47.1	44.4	54.1	80.9	0.49	0/9
169 Ethiopia	149 0.393	53.1	50.5	22.8	50.0	0.8	1.3
173 Mali	151 0.371	55.3	50.8	15.9	32.7	0.8	1.2
176 Burkina Faso	154 0.364	52.9	49.8	16.6	31.4	1.0	1.4
177 Sierra Leone (lowest ranked)	157 0.320	43.4	40.2	24.2	46.7	0.5	1.1

(after UNDP Human Development Report 2007–08)

Gender concentrations

Park and Burgess (the same person who identified the model) noted in the 1920s what they called **moral regions,** the areas of prostitution and vice in Chicago. In San Francisco, Castro is the district where gay men live and male prostitution is found on Polk Street. Female prostitution is found at Mission and 18th Street. Valencia Street is the core of the lesbian community. Can the same be found in UK cities? Canal Street in Manchester is associated with the gay community, but as a recreational area rather than where they live.

There are areas of cities that can contain more of one gender. In some cities the nature of the housing market and the type of housing enables single females to purchase properties in an area. This has happened in Central Southsea where the ward has a disproportionate number of households occupied by 20–39-year-old single females.

1.6 What are the demographic challenges facing countries?

Changing age structures

Figure 18 shows how age structures have changed in MEDCs and LEDCs from 1950 to 2000.

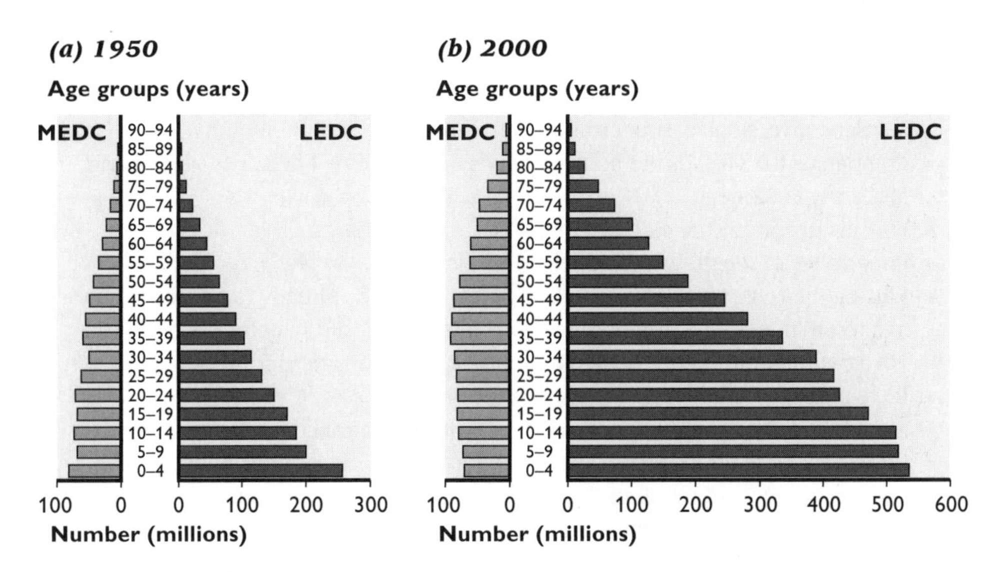

Figure 18 Age structure of MEDCs and LEDCs: a) 1950, b) 2000

Impacts and issues in MEDCs

In 1901 in the UK, life expectancy for men was 45 and for women was 49. In 2001, it was 75 for men and 80 for women. Women aged 65 in 2001 could expect to live until they were 84, whereas men of the same age could expect to live until they were 81. In 1998, the over 60s outnumbered the under 16s for the first time. The proportion of the elderly in the UK is rising faster than anywhere else in Europe.

1 Ageing society

In 2006, 16% of the UK population were over 65, those over 85 were 5.9% of the population and by 2031 the figure for the over 65s is forecast to be 22%. When you retire you will be part of the 29% over 65. The elderly are dominated by females with a ratio of 234 females to every 100 men among the over 80s. The baby boom children of the post-1945 period are now reaching retirement. The elderly often migrate on or just before retirement — **retirement migration.** This may be caused by **downsizing** to a smaller property, or the desire to live in one of the retirement regions of the UK such as the southwest, coastal East Anglia and south coast resorts. Overseas retirement migration has grown although it may be slowed by the 2008–09 credit crunch's effects on private pensions. Some concentrations of older persons are found in the inner city where they have been trapped by low incomes and the dependence on social services and pension credits. Out-migration of the young from areas such as mid-Wales has also led to an apparent concentration of the elderly. Some concentrations are caused by **return migration** as those who left to work in the cities 40 years ago now return to their roots.

The effects of an ageing society on policies can take many forms:

- The conversion of larger properties to nursing and care homes. New purpose-built facilities may also be developed. However, some local authorities are seeking to restrict conversions of properties to care homes, e.g. Bournemouth.
- Companies to cater for the housing needs of the elderly have been established, e.g. McCarthy & Stone.
- Policies to employ the elderly, e.g. B&Q.
- Importance of the grey vote, reflected e.g. in policies on law and order.
- Raising the retirement age — the UK government has already raised the retirement age to 68 in 2044 (i.e. for all those born after 1976) and may increase it further. In 2010 the minimum retirement age for private pensions rises from 50 to 55.
- Increased leisure facilities, e.g. Saga holidays.
- More businesses geared to the elderly, e.g. RIAS insurance for the over 50s.
- Pressures on health services ranging from care to geriatric beds in hospitals.
- Local economy and retailing changes to cater for an elderly population.
- Growth of new educational provision, e.g. University of the Third Age.

Conversely:

- Fewer people will be moving into work.
- There will be less demand for teachers and schools.
- There will be fewer people in work to support the dependent elderly.

Policies to alleviate demographic challenges

Expansionist policies

- Make abortion illegal, e.g. Guernsey.
- Support for working mothers, e.g. France.
- Improve health care, e.g. UK.
- Restrict availability of birth control, e.g. Ireland and Ghana.
- Give larger family allowances to those with more children, e.g. France today and USSR in the twentieth century.
- Restrict the role of women in society, e.g. Saudi Arabia, and Afghanistan under the Taliban.

Control policies

- Promotion of birth control, e.g. Singapore.
- Health education in schools, e.g. Netherlands.
- Control family size by controlling number of births per family, e.g. China's one child policy (not being enforced everywhere in China in recent years).
- Permit abortion, e.g. UK.
- Encourage sterilisation, e.g. India.

2 Singapore, growth in a confined space

Singapore has a population of 4 million people concentrated in a small area, with no natural resources other than the ability of its people. Singapore's policy is to encourage parents to have only one child. To provide the wealth necessary to support the growing population and to improve living standards, the state has invested in high technology and has created a financial centre for the region. It has built a science park and a major conference centre.

Singapore encourages in-migration of skilled labour from Europe to supplement the local workforce and enhance the profitability of the service sector. These activities earn more money to support the building of new towns, for example Tampines, to house the population. More recently, the state has sponsored industrial investments overseas in the Indonesian Riau Islands, Malaysia (Johor) and China to gain profits by being the regional headquarters for overseas workbench assembly lines. Low-skilled, low-profit activities have been relocated out of the country so that high-skilled, high-profit activities can take place in Singapore.

The Singapore government has invested heavily in education and has encouraged its educational system to invest overseas. It has developed a global airline (Singapore Airlines) to provide foreign earnings from its activities, which include a 49% stake in Virgin Atlantic. All of these efforts are underpinned by strong state planning using 5 year plans. As a result, Singapore has developed a resource base to support and sustain moderate population growth.

3 India: coping with growth

India has made progress to reduce its population, yet it is still ranked 126th out of 177 countries on the UN Human Development Index (HDI). The birth rate was 41/000 in

1951 and had fallen to 25/000 in 2005. At the same time the death rate declined from 23/000 to 8/000. In 2001 34% of the population was below the **poverty line** (defined as being able to receive 2,400 calories per day in rural areas and 2,100 in urban areas). Kerala has only 10% poverty whereas Bihar and Rajasthan have over 50%.

Family planning (now called **family welfare**) is encouraged, as is sterilisation (at its peak in 1976 there were 8.3 million sterilisations performed). Abortion has been permitted since 1972. In addition, **feticide,** the abortion of girl fetuses, has been common among the middle and upper social groups in richer regions such as Delhi. Female infanticide, because of dowry issues, is still practised in Rajasthan. Therefore, India has a **gender imbalance** irrespective of population policies. Education, particularly of women (illiteracy is 52% among women and 25% among men), will assist in lowering the birth rate. There are travelling contraceptive clinics in rural areas.

Other population policies have focused on feeding schoolchildren (53% of children aged 5 are underweight), and building rural homes to raise standards of living. Grain production quadrupled in the late twentieth century thanks to the **green revolution** and better health control. Improved health provision halved infant mortality.

India is also following a path of developing its manufacturing and service sector (call centres and computer software houses) in order that it improves its resource base as well as controlling population growth. The task is still daunting if the decision to import cereals in 2009 is a sign of the growing population's demand for food.

4 Botswana: coping with decline in a LEDC

Botswana is a small, landlocked semi-desert country in southern Africa. In 2005 it had a population of 1.8 million. It has a maternal mortality rate of 380/000 live births, which has risen in the past decade. The under-5s mortality rate is 120/000 and there are 40 doctors for every 100,000 inhabitants. Botswana's population would grow more rapidly but for the ravages of HIV/AIDS. Life expectancy is 48 years because of AIDS related deaths. At one stage, the UNDP forecast that life expectancy would decline further, but this has not materialised. HIV is prevalent among between 23% and 32% of all people aged 15–49. The spread is closely linked to poverty, particularly because of the victims' lack of education, information and access to health care. Women are more vulnerable if they are poor and in a society tolerant of extramarital sex, where partners are promiscuous.

5 Uganda

In 2005, Uganda's population was 26.9 million. Growth is slowing and is currently at 2.4%. There is a maternal mortality rate of 6.9/000 live births. The mortality rate for under-5s ranges from 105–194/000 depending on the income of the family. Illiteracy stands at 38% and there are 8 doctors for every 100,000 inhabitants. HIV/AIDS has had a similar effect to that in Botswana although it has declined markedly to between 5.7% and 7.6% of those aged 15–49. The government has recognised that the issue is developmental and that it will take time to change attitudes. It has begun campaigning against high-risk behaviour, empowering communities through the work of NGOs, educating employers to assist, and providing better access to health care. Today, only 8.3% of adults have HIV, because of the success of these measures over the past decade.

6 Thailand: community-based policy

In 1969, women had, on average, 6.5 children and only 16% used contraceptives. This resulted in a population growth of 3% per annum. The solution was a nationwide family planning programme which commenced in 1970 and included free contraception, trained family planning specialists, and government campaigns, especially in rural communities. By 2005, 79% of married women were using contraceptives, women averaged 1.7 children and population growth was 0.8% per annum. The population will continue to rise but the slower rate of increase is the result of the successful community-based rather than coercive policy.

Theme 2: Investigating settlement change in MEDCs

2.1 The features of settlements

The rural–urban continuum

Rural and urban environments are closely linked to each other, for example, through food supply, and work or leisure activities. It is very difficult to distinguish between rural and urban environments using only one feature. In reality there is a **rural–urban continuum.** This topic looks at the criteria that distinguish rural and urban areas.

Size

Many countries employ a **population threshold,** for example 20,000 people, for a settlement to be classed as urban, but the threshold value varies between countries and with the character of the area. In sparsely populated areas, such as western Canada or northern Sweden, a settlement of 1000 people may be more urban in character than an agro town of 30,000 people in southern Italy. The range of minimum populations is from 200 in Norway to 50,000 in Japan. For a UK parish to be classed as rural, the Countryside Commission suggests it should contain fewer than 10,000 people.

Function

Urban areas have a wide range of functions, such as cultural activities, retailing, education, administration and financial services. These functions are of a higher order (level) in urban areas, for example, a department store or a college as opposed to a village store or primary school. Urban areas, therefore, are central places and they have a sphere of influence. In contrast, rural areas usually have comparatively few services and most of these are concentrated in small market towns.

Employment

Traditionally, the main source of employment in urban areas is either in the secondary (manufacturing) sector or in the tertiary (services) and quaternary (research and administration) sectors. Urban areas attract employees from their catchment areas.

There is little, if any, primary activity in urban areas in MEDCs. In MEDCs in rural areas less than 10% may work in the primary sector (it can be as low as 2%).

Density

Urban areas are built up. **Threshold population densities** can be used as a measure of this built-up nature. The usual range is from 100–400 people per km^2. Rural areas in MEDCs are frequently sparsely populated although there can be quite high densities around major cities.

Administration

Many urban areas function as centres of government with some centres being responsible for several layers of administration.

Character

People like to suggest that urban areas have a social dimension — an urban lifestyle (**urbanism**) that is characterised by pace, stress and a desire to be upwardly mobile. Urban environments are also characterised by a more obvious polarisation from extreme wealth to acute poverty within small areas. People perceive rural environments as stress free, with a slow pace of life — a rural idyll. However, rural environments in MEDCs do contain pockets of poverty.

The **rural–urban continuum** covers a **hierarchy of settlements:**

- **isolated dwelling**
- **farm**
- **hamlet** — a small cluster of dwellings/farms that lack services
- **village** — a settlement in rural surroundings that was once an agricultural settlement. Villages normally contain some services such as a church, an inn and possibly a shop. The term village can be used by estate agents to try and enhance the identity of settlements and price of property in former villages engulfed by suburbia
- **large village** — has more services such as antique shops and mini-mart retailing
- **town** — smaller urban area with a range of facilities to serve the area: post office, schools, a range of mainly independent shops
- **city** — a large settlement depending on commerce, manufacturing and service industries. The city region is the area served by a city and usually includes the journey to work area and, importantly these days, the journey to study area
- **conurbation** — urban areas which gradually fused together in the nineteenth century such as Greater Manchester and Leeds–Bradford and the Ruhr area of Germany
- **mega conurbation** — a merging of conurbations whose sheer size is very extensive such as Boswash, the US seaboard from Boston to Washington, and SanSan, coastal California from San Francisco to San Diego
- **megacities** or **global hubs** — the key centres of the global economy: London (11.9 million people), New York (21.8 million people) and Tokyo (34.1 million people) are the centres of banking, finance and investment (the credit crunch may have diminished their reputations but they remain the key cities). In the future, Shanghai (17.9 million people) and Singapore may have this status (but not population size in Singapore's case)

Millionaire cities in population, are somewhat different from the trio of global hubs: Mexico City (22.6 m), Seoul (22.2 m), Sao Paulo (20.2 m), Mumbai (19.7 m), Delhi (19.5 m), and Jakarta (17.1 m). Other than the top trio, Los Angeles (17.9 m), Osaka (16.8 m) and Moscow (13.7 m) are the only megacities in population in the world's top 20.

All characteristics become more extreme the higher the position in the hierarchy. For example, a village might offer a primary school and a pub while a city might offer specialist cultural facilities, a university and a number of regional offices. These high-order functions give the city a very large sphere of influence. Obviously, the higher up the hierarchy the fewer the number of settlements.

Urban environments

Urbanisation is the process by which the proportion of the population living in an urban area increases. It is a process of change that affects both the places themselves and the people involved. It is also a multi-strand process — over time all the urban characteristics identified above occur. A number of changes result as urbanisation occurs:

- A **sectoral shift** in the economy of a region or a country with an emphasis on moving from farming in the primary sector to manufacturing and the provision of centralised services which is linked to economic development.
- **Urbanism** — as people move to towns and cities either from rural areas or from other urban places, they change occupations as they strive to improve themselves. Urban dwellers encounter high costs of living and commute long distances. Inevitably, a substantial minority do not succeed and an underclass develops.
- The population **distribution** changes. A combination of rural to urban migration and the resultant natural increase from young migrants can result in rapid population increases in cities. Conversely, population loss from rural areas and an ageing population slows down the rate of rural population change. Some movement back to rural areas (**counter-urbanisation**) will also boost rural populations within commuting distance of urban areas.
- Changes within the settlement **hierarchy**. The number of urban areas grows rapidly as urbanisation occurs (e.g. nineteenth century UK). Small towns grow and larger ones coalesce to form conurbations.

Urbanisation over time

There is a very close relationship between economic development and urbanisation. Figure 19 shows a model of the process that has been followed by MEDCs over time. (NB you only need to know the process as it applies to MEDCs.)

In the eighteenth century most MEDCs, such as the UK and Germany, experienced the Industrial Revolution, based on coalfields with extensive manufacturing industries. It took about 150 years for urban percentage of the population to rise from 20% to 75%. Rates of urbanisation were slow because of high death rates with low life expectancy and very high infant mortality.

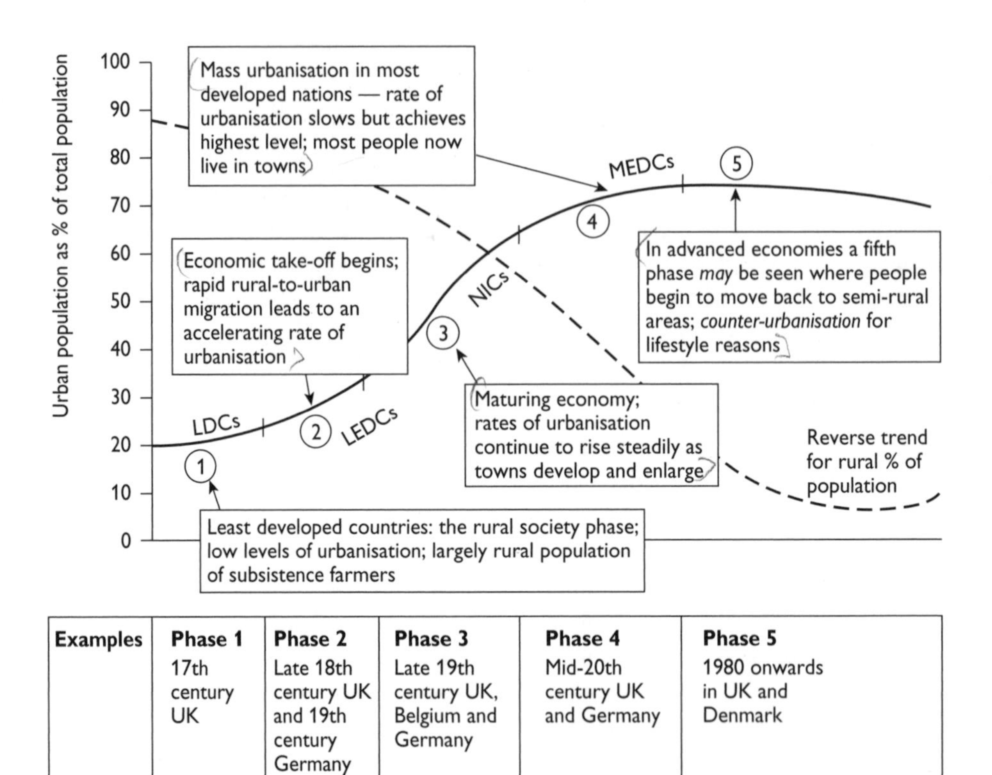

Examples	Phase 1	Phase 2	Phase 3	Phase 4	Phase 5
	17th century UK	Late 18th century UK and 19th century Germany	Late 19th century UK, Belgium and Germany	Mid-20th century UK and Germany	1980 onwards in UK and Denmark

Figure 19 The process of urbanisation

In the twentieth century, cities consolidated their hold on economies and developed a wide range of activities that employed an increasingly diverse population. By the end of the century, in advanced economies people were beginning to move back to semi-rural areas — **counter-urbanisation** — or back into cities **reurbanisation.**

The cycle of urbanisation

The present balance of centripetal and centrifugal forces shown in Figure 20 changes over time, either in response to commercial forces or in response to government planning decisions. The interplay of these forces has a major impact on the cycle of urbanisation and the evolution of the city structure and shape.

Perceiving places

Throughout this book you will see references to how people perceive places. What people refer to as a village can vary. Most see a village in line with our definition but it is used particularly by estate agents to suggest a small community even if that community is engulfed in suburbs. Is Greenwich Village on Manhattan, New York a village or is it merely an area with distinct social and physical characteristics? Is the term 'market town' really describing the town with a rural market or is it trying to convey a small feel to a settlement that has expanded rapidly over the past 30 years?

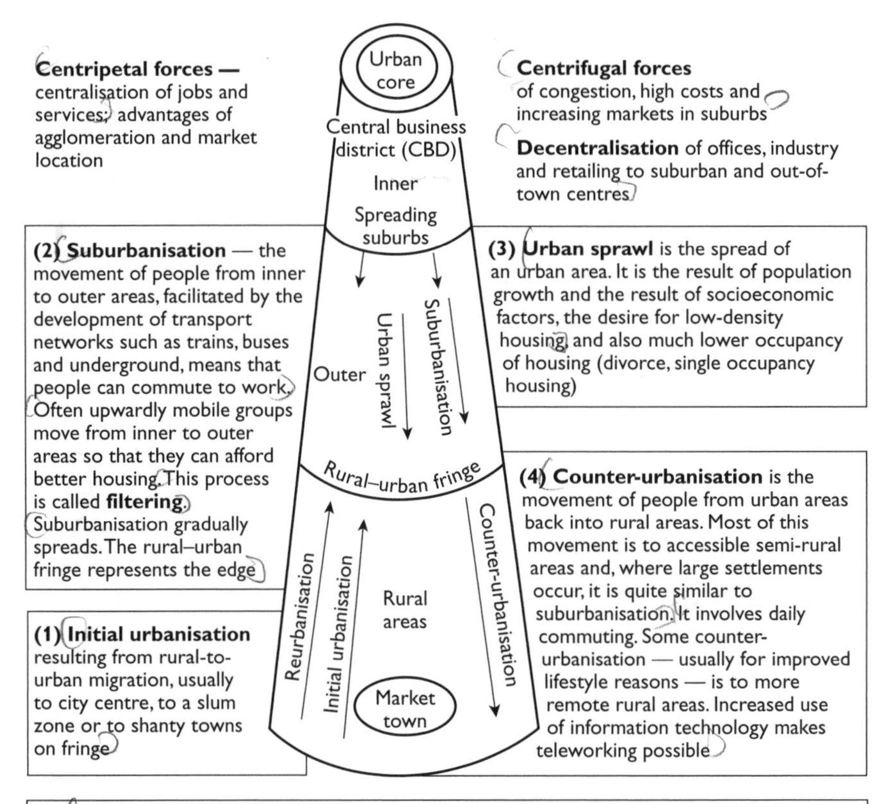

(5) Reurbanisation involves a range of processes which enable people and economic activities to move back to city centres. **Gentrification** is when middle-class people move back to run-down inner urban areas and improve the housing stock. Some reurbanisation results from planned initiatives such as those from Urban Development Corporations where inner central areas are improved in a number of ways, with high-value housing, hi-tech employment and improved environments

The cycle of urbanisation
- (1–5) is the typical sequence of processes in an MEDC.
- The processes result from the balance of centrifugal and centripetal forces. If there is outward movement, centrifugal forces are dominant.
- In some MEDC cities centrifugal forces have become so strong that planners are working to regenerate the centripetal forces

Figure 20 The cycle of urbanisation

2.2 How does the social and cultural structure of settlements vary and why?

Urban morphology

Urban morphology can be defined as the form or internal structure of an urban area or how the various functions found within a town or city are arranged. Often distinct land-use **zones** can be recognised. The main reason for zoning is the cost of land. **Bid rent theory** is often used to explain zoning. It states that all functions have a price they are willing to pay for a piece of land in an urban area. For example, in the city centre, land prices are very high. Only land uses such as prestigious financial institutions and/or quality retailing can afford these costs (purchase price or rent and council tax) and so they are able to outbid other uses such as residences.

- **Agglomeration** is also a significant factor in zoning because many businesses benefit by grouping together within a business/office zone. Increasingly, positions at major road intersections are highly prized and valued by functions requiring maximum access to consumers — e.g. car showrooms, often called **motor rows**.
- **History** is a very significant factor in influencing how a settlement developed. Urban areas in MEDCs generally developed from a historic core and spread outwards in a concentric fashion — so housing is generally older near the centre and much of it can be of poorer quality compared with the newer housing in outer suburbs. Sometimes historic buildings such as castles and cathedrals can distort the pattern by preventing spread in one direction or retaining an ecclesiastical quarter.
- **Specialist site factors** influence morphology in a particular way. Many cities have a waterfront industrial area but the function has changed as the area became derelict. This has provided opportunities for **regeneration** (e.g. Cardiff Bay, Swansea SA1 and Albert Dock, Liverpool).
- **Political factors** (both central and local government and in the case of Wales, the Welsh Assembly) play a major part in how land is developed or redeveloped. City planning departments have strict laws on zoning — controlling where developments can take place.

TIP You need to be able to draw a sketch map to show the morphology of at least one MEDC city (see Figure 21).

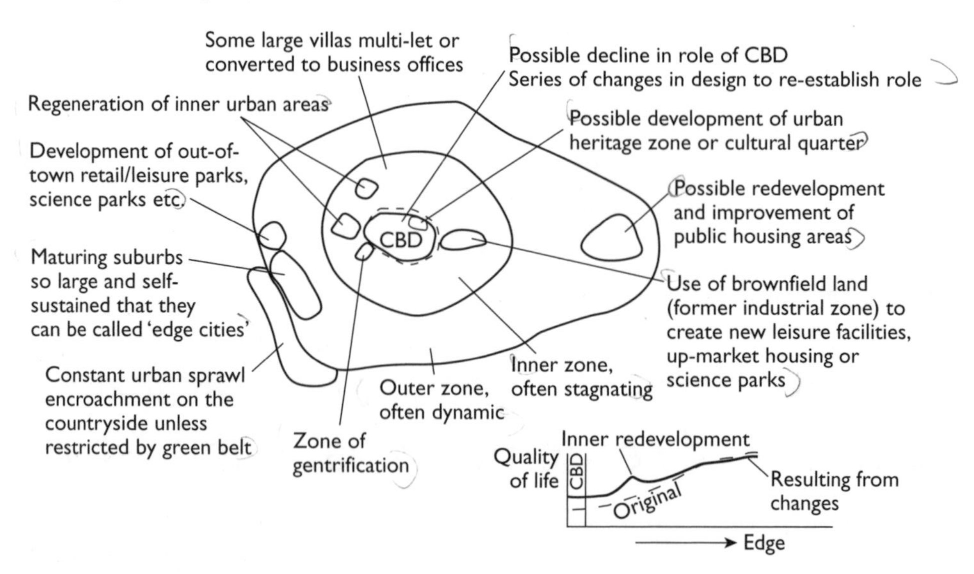

Figure 21 Changes that may be occurring in an MEDC city

Urban areas are constantly changing. Figure 21 provides a checklist for the changing morphology of a city in an MEDC. Adapt this diagram for your own examples, annotating it to describe land uses and identify the factors responsible. While you may find many of the developments shown, some will be lacking in your chosen example.

Urban models

An **urban model** is a simplified diagram created by looking at a range of cities and making generalisations, to include similar features. A combination of fieldwork (primary data collection) and map and census analysis (secondary research) can be used. Once the model has been drawn it is possible to compare it with any city or town and you should try to explain why the morphology of your chosen city or town is similar or different. Always use a model that is appropriate for your chosen city. Some models such as Burgess' concentric rings were developed over 89 years ago for completely different circumstances (very rapid growth) from those experienced in modern MEDC cities. Other models, such as the multiple-nuclei, only apply to cities that developed from several centres, e.g. Los Angeles or Stoke-on-Trent. For studying UK cities, especially the northern industrial cities, Mann's model developed in the 1960s is useful (see Figure 22). Another UK model that you might research is Robson's, based on Sunderland.

A Middle-class sector
B Lower middle-class sector
C Working-class sector
(and main sector of council estates)
D Industry and lowest working-class sector

1 Central business district
2 Transitional zone
3 Zone of small terraced houses in sectors C and D; larger housing in sector B; large old houses in sector A (Victorian)
4 Post-1918 residential areas, with post-1945 housing on the periphery
5 Commuter villages (post-1960)

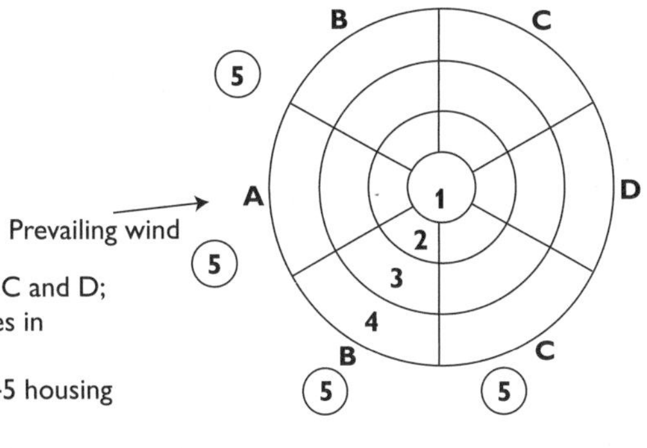

Prevailing wind

Figure 22 Mann's urban structure model

Inequalities and social problems

Most urban areas consist of a mosaic of neighbourhoods of differing quality of life, but underclasses tend to be concentrated in the very worst physical environments of an urban area — usually because they have no choice. The richer inhabitants tend to encourage the development of exclusive neighbourhoods so their paths do not cross with those of poorer inhabitants. Separation of rich and poor can become very marked, especially as the number of cyclical processes leads to a downward spiral of **multiple deprivation** in poor estates. And so problem areas develop (see Figure 23).

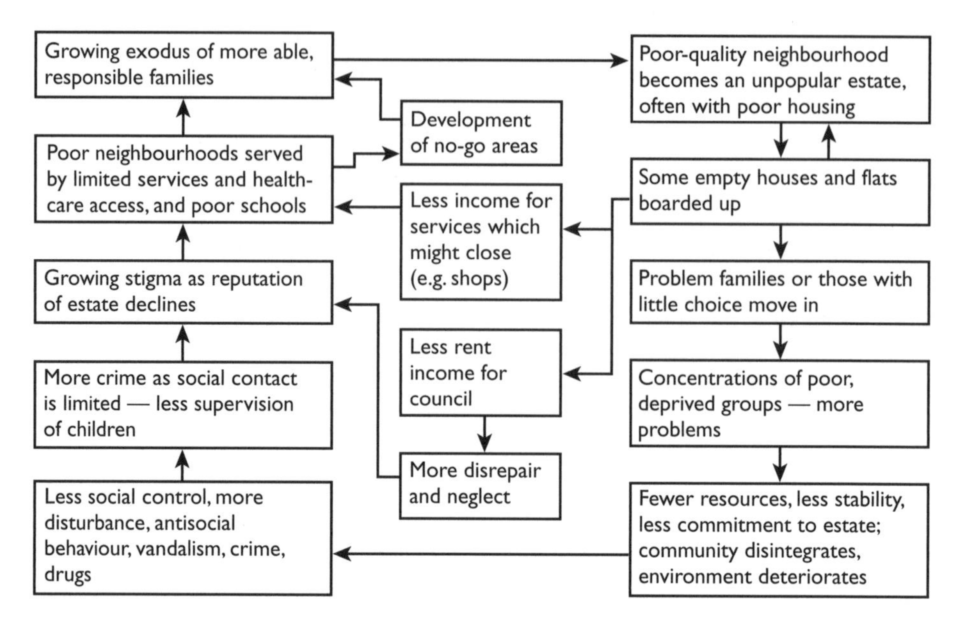

Figure 23 Spiral of multiple deprivation on poor estates

Cities are noted for their **multi-cultural plurality**, which can lead to a mixing of ethnicities, religions and cultures. It can lead to a mosaic of **ghettos**. A ghetto is an area where the population is almost exclusively made up of a single ethnic or cultural minority, e.g. Black and Puerto Rican Harlem in New York. Ghettos are common in American cities and are increasingly found in European cities. They develop especially in times of unrest that lead to fear and the need to cluster. They also develop when new immigrants from different cultures enter a city, e.g. Turks in Köln, Germany.

Ghettos are frequently concentrations of deprivation. Many new immigrants arrive with limited resources, and discrimination can cause higher levels of unemployment and lower incomes among these groups. As ethnic groups are forced to concentrate in areas of poor housing, this pushes out groups from other races (called **white flight** in the USA). Levels of crime and unrest are frequently higher — hence the turbulence of places like south central Los Angeles. Figure 24 shows the factors that encourage ethnic segregation. In Milan, the Roma (Gypsy underclass of illegal and genuine immigrants) are housed in makeshift camps or under railway arches and are subject to forcible removal by the authorities — but they return.

Ethnic clustering is a complex topic and patterns of ethnic groups vary. In cities where immigrants are long established, well integrated, and not threatened by racial discrimination, the groups tend to disperse throughout the city as their confidence and affluence increases. This has happened in Sydney and Melbourne and is known as **mixing**. Multinational plurality prevails in many Australian cities.

Internal factors — those within ethnic groups that encourage segregation

New arrivals need mutual support from friends, relatives and community organisations

These immigrants will be happiest with religious centres, ethnic shops and foods, and banks grouping together to serve them

These immigrants need support from areas speaking their own language — a minority language in their new country

These ethnic groups encourage friendship and marriages and reduce contacts, except via schools, which might undermine the culture and traditions of the ethnic groups

Jobs and accommodation can often be obtained via networking in an ethnic community

A closely knit ethnic community provides back-up against abuse and racist attacks

Ethnic groupings help political power and influence development

External factors — those within the country or area that encourage ethnic segregation

As immigrants move in, the remaining majority population moves out — in fear of factors such as falling house prices

The majority population is generally hostile or unfriendly to new arrivals

Racism, abuse, racially motivated violence against ethnic minorities or fear of such actions

Discrimination in the job market — ethnic minorities in low-paid jobs or unemployed are forced by circumstances into cheap housing areas and substandard services

Discrimination by house sellers, estate agents and housing agencies keeps ethnic minorities in their ghettos

Discrimination by financial institutions forces ethnic minorities to use their own networks for small business development etc.

Ethnic segregation in urban areas leading to concentration

Figure 24 Factors that encourage ethnic segregation

- Some concentrations occur in the central area, such as the African-American ghetto in Watts in central Los Angeles. Sometimes this has been for historic reasons where there were large quantities of cheap housing to rent or to buy. In other cases, concentrations are in poor quality housing in the suburbs.
- Recent immigrants, such as Mexicans, have concentrated in barrios in, for example, east LA and San Fernando. As they secure better jobs and learn the language they may leave these barrios but will not do so if they feel racially threatened.
- Immigrants also live in what are called the ethnoburbs of the Pacific Rim: middle-class or professional immigrants settle in affluent suburbs where they invest in businesses and develop properties. These new clusters form due to choice, convenience and cultural reasons and are often influential economic enclaves.
- Elite ghettos can occur in exclusive neighbourhoods such as the new gated communities in Beverly Hills. This type of ghetto was common in colonial settlement where the colonial rulers were physically separated from the local people (sometimes these areas remain as enclaves, e.g. Victoria Peak in Hong Kong).

Most of the distribution outlined above results from natural processes, but under extreme racial or political tensions — as in parts of Northern Ireland or Bosnia — the fear and reality of persecution and extreme danger can lead to systematic **ethnic cleansing** or the withdrawal of various groups into increasingly 'pure' neighbourhoods, sometimes physically separated by a wall.

Ethnic concentrations are a major cause for concern when ethnic minorities become an underclass concentrated in the least attractive, socially excluded estates. In MEDC cities these are commonly rundown inner-city areas, but they can equally be problematic estates of **social** or **welfare** housing either in the inner city (Hulme in Manchester) or on the city outskirts (Chorweiler, outside Köln, Germany).

The changing employment market has affected both types of area, with many low-paid, part-time, insecure jobs available. The cycle of disadvantage and deprivation is very difficult to break as it tends to be passed from generation to generation — a fact reinforced by poor health and educational facilities in these areas. It can also be reinforced by racial and other forms of discrimination.

Solutions

A number of controversies exist as to how to best solve problems of inequality:

- Should the finance come from public or private sources or a mix of both? The 2008/09 credit crunch had its origins in private finance for loans in poor areas that led to toxic debt.
- Should the initiatives come from central government (top down) or via local community-based projects?
- Is tackling the environmental, the social, the economic or the political problems the key priority?
- What is the best way to spend any money allocated — saturation in one concentrated area or dispersed initiatives?

Most cities have their own websites detailing their schemes.

Tip: always check the bias of the sources — the official statements on the success of improvement schemes are often more optimistic than the views of the people living there.

Studentification

A more recent phenomenon of segregation is the development of student districts in university towns. Once these were the colleges of Oxford and Cambridge, but today large areas of inner-city wards near universities have become residential districts for students, e.g. Southsea for the University of Portsmouth.

Perceptions of urban structure

Most people carry a mental map of an urban area based on their experience and education, as well as the further experiences they receive as they journey about the area. Lynch modelled the way people see cities:

- **Paths** — our channels of movement along roads or rail lines or where we walk or what we will see as we use the images in Streetview or from a sat nav.
- **Edges** — the barriers that cut areas off from one another. They can be urban motorways, rail lines, a river, coastline, or even the boundary of a gated community. Some edges do become seams that bind areas together, such as the Cam in Cambridge, which binds the colleges into a whole.

- **Nodes** — the focal points in a town or city where major transport pathways meet or where there is a concentration of activity. The area just outside many rail stations is a node in most people's minds. Canary Wharf is a node as is the Welsh Assembly building in Cardiff.
- **Districts** — areas identified by a common characteristic, normally its architecture. Georgian Bath is an easily perceived district as are many city centres, although these may be perceived in part rather than as a totality.
- **Landmarks** — the features with distinctive characteristics that people remember. They can be very local, distant landmarks or even outside the urban area.

How people see all of the above depends on:

- Their age — a 3-year-old child may see the 'dong clock' — Big Ben — but not notice the Thames. Young adults might see more in an entertainment district than their parents who perceive it differently.
- Gender.
- Experience and frequency of visiting an environment — familiarity aids recall.
- The social class of the individual or group — for example, are 'no-go' areas perceived as a result of social class or perhaps they are areas where housing is not for sale and only available for rent.
- The value system of the individual or group. Many ethnic and religious districts grew up because they were the areas where a person's value system was best appreciated, e.g. the Jewish area around Golders Green.

There are many possibilities for field study of perceptions of a city or a village.

2.3 What are the issues of the inner city?

Inequalities

Inequalities are found in all urban areas and the scale of the inequalities reflects national and regional patterns. Urban models show that contrasts in wealth can be found over very short distances, which should be noted in your field studies of a city. Both the wealthy and the poor concentrate spatially and there are reasons for this **social segregation** in the inner city and beyond.

The key is housing, which is fixed in location. Developers and planners tend to build housing with a particular market in mind. Wealthier groups can choose where they live — paying premium prices for areas well away from poor areas, with pleasing environments and quality services, such as schools. The poorer groups have fewer choices and are, in a sense, forced to live where they are placed — in social housing or where they can find a cheap place to rent privately.

Housing needs to be considered in the context of changing environments. Housing neighbourhoods change over time — large Victorian villas are now too large for the average family and so have been 'converted' into apartments for private renting. Other neighbourhoods have improved because of **gentrification**. The same type of housing can be built in different environments. It may be on an attractive hillside

(Hampstead) or in a recently improved neighbourhood targeting upwardly mobile people. **Right to buy** transformed some council estates so that, for example, rows of houses facing fields were bought and improved. As we have seen, immigration can result in inequalities.

Measuring inequalities

It is possible to measure **quality of life** in an area, such as the inner city, using **primary** data. This data includes quality, density, and condition of housing, and the nature of the environment (physical and social). Levels of pollution or crime are good indicators here (Figure 25).

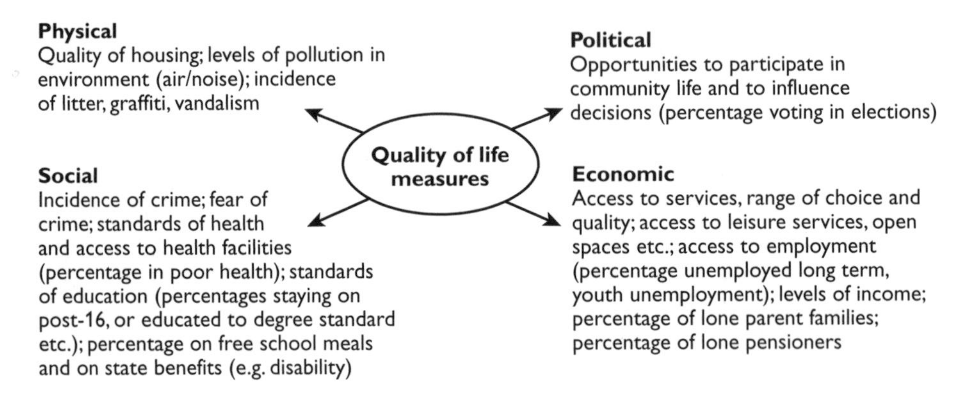

Figure 25 Quality of life measures

It is possible to use **secondary** data from a census to assess **deprivation** levels — this may include **poverty** in terms of low income or can be manifested in poor health or lack of possessions, e.g. a car. It is very common for the poorest parts of an urban area to suffer from **multiple deprivation** — with social, environmental and economic deprivation. There have been many indices devised to measure deprivation, including Townsend's. (It would be useful to research the Townsend Index of Deprivation and Disadvantage.)

Urban social exclusion is a term that refers to the problems faced by residents in areas of multiple deprivation, such as the inner city. Essentially, these people are excluded from full participation in society by their social and physical circumstances. They cannot get access to a decent job because of poor education, or obtain decent housing because of poverty, and have to suffer high levels of crime and poor health in an unattractive physical environment.

Inequality can have major consequences in a city in terms of a lack of social cohesion. In extreme cases it can lead to civil unrest. Governments have to address social injustices for a variety of social, economic and political reasons. Figure 26 shows an idealised quality of life transect across a city in the UK and the lower quality of life in the inner city.

Urban issues are summarised in Table 7. Many of these are wider in extent than just for the inner city although they are measurably more intense in the older parts of a city.

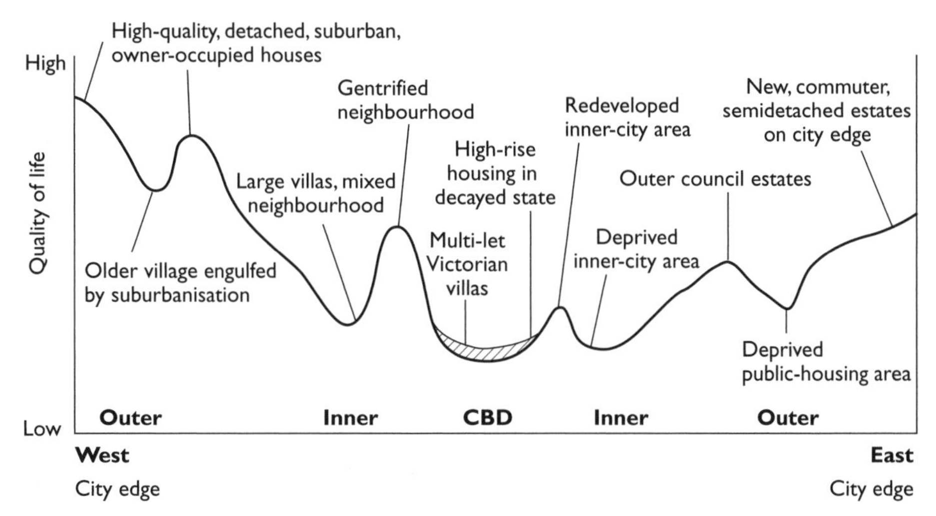

Figure 26 Idealised quality of life transect across a UK city

Table 7 An overview of urban issues

Environmental	• Problems caused by atmospheric pollution from a variety of sources, especially from industrial, domestic and transport services • Problems of traffic congestion – moving the daily flow of commuters • Problems caused by sewage and waste disposal • Issues of water pollution • Problems generated by rapid urban sprawl, in particular loss of land and creation of wide areas of slums • Dereliction may be a problem, with many 'toxic' or unsightly areas • Some cities have specialist problems such as flooding or landslides – particularly problems for poor people who cannot choose to live in safe locations
Social–cultural	• The development of zones of poverty within the city – can become areas of high crime and social deprivation • Ghettoisation of various income and ethnic groups is leading to friction between groups • Creation of an underclass of people with limited powers to challenge their poor quality of life and health
Economic	• The problems of providing for the enomous numbers of new arrivals • Services such as health, education, housing, employment and utilities such as water and sewage are increasingly costly to provide
Political	• The difficulties of governing mega-cities effectively to cope with the enormous range of problems; declining incomes from rates

Housing stock

Housing problems are some of the most widespread and serious of all the troubles facing inner cities. They have many facets:

- The quality of housing — size, quality of construction, provision of services such as water, electricity and gas, and the safety of the site. Poorly constructed houses that are overcrowded have a major impact on the health and well-being of residents. For example, consider many of the problems associated with high-rise blocks in the inner city.
- The quantity of housing relative to the number of households that require homes. Often the demand for housing exceeds supply. This has a knock-on effect on other factors, such as finding workers. Where there is a shortage of housing, some people are inevitably housed inappropriately.
- The availability and accessibility of housing — the ease with which people are able to buy or rent or self-build housing at affordable prices. The right to buy policy removed large quantities of inner-city housing (and outer city estates) from the renting sector. In many cities prices are rising and people are unable to get onto the private housing ladder. This is why affordable social housing in MEDC cities is of vital importance in solving housing problems — especially in the inner city.
- The range of housing tenure — the housing mix. A good balance between the supply of owner occupied, privately rented and social housing in relation to demand is essential in all cities.

The downward spiral of housing for the urban poor

While the rich have the opportunity to choose where to live, providing decent housing for the poorer population of urban societies is a major challenge. Low incomes mean little or no choice — being forced to rent, with little security of tenure or protection against sudden or forced eviction. Poor people have to put up with inadequate housing, for instance in tower blocks or overcrowded apartments. The houses are normally in unsafe locations, often perceived as problem areas, which prevent their inhabitants from access to credit. In many cities, in spite of attempts by national and local governments, the problems are getting worse rather than better because decent, well-located yet affordable housing is in very short supply. Many people have to travel long distances to work in order to obtain reasonable housing. As can be seen in Figure 27, there can be many areas of poor housing that are not located in the inner city.

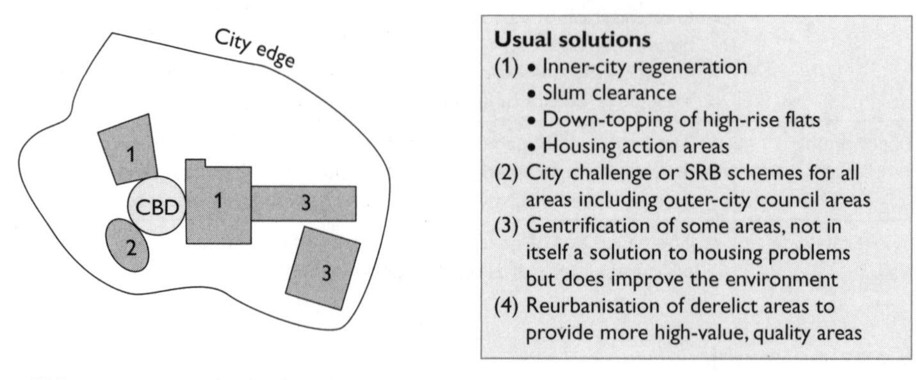

Figure 27 Areas of poor housing in an MEDC city

One response to the strategy of developing appropriate housing within the inner city is **gentrification** (Figure 28). This is the process by which relatively affluent middle-class households move into run-down, inner-city neighbourhoods. It applies to small areas in a neighbourhood, but often has knock-on effects and steadily spreads. Some planned gentrification of old warehouses in docklands is on a very large scale, e.g. the Wapping area of London Docklands.

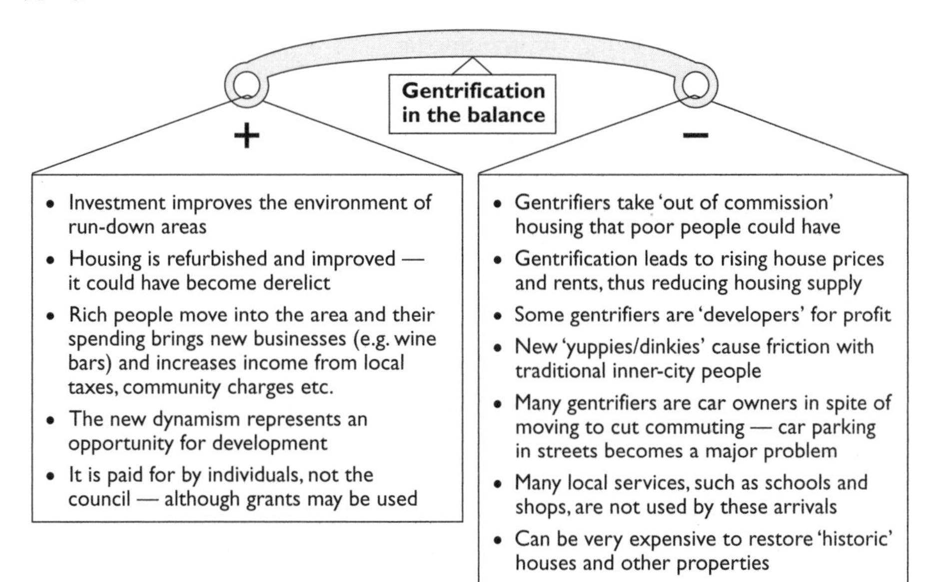

Figure 28 Gentrification in the balance

Gentrification is one component in the **reurbanisation** process that is taking place in the inner city to combat urban sprawl and over development of **greenfield** sites. Another strategy is to build new-style housing — mainly for young executives — on **brownfield** sites, or to refurbish old industrial buildings to provide luxury apartments. Sometimes, as in London Docklands or Newcastle-upon-Tyne, inner-city nineteenth century terraced housing has been knocked down displacing the existing population. Here, large-scale development of upmarket housing is seen as part of the overall process of regeneration and environmental improvement. The displacement of poorer people from large parts of the inner city, when in fact they form a vital component of the workforce that keeps a city running, is a major issue in MEDC cities. Displacement, combined with changed family circumstances, can generate the first stage in the cycle of homelessness.

Regenerating and redeveloping urban areas

Figure 29 summarises the major issues of regeneration.

Regeneration can lead to a number of possible improvements:

- Dealing with social problems such as poor health, low educational standards or substandard housing in **action zones**. These programmes tend to be community based, although they can be top-down in design.

- Targeting economic improvements that relieve poverty and thus solve some social problems or provide new employment, e.g. **urban tourism** in Bradford.
- Improving the environment, for example turning derelict waterfront properties into an attractive environment provides a catalyst for development. This has been particularly successful in Barcelona and many other former dockland areas.
- **Image enhancement** or **urban boosterism** — by promotional programmes to change outsiders' perceptions (vital to attract inward investment). Often based around promoting cultural and sporting facilities, e.g. the 2002 Commonwealth Games in east Manchester, the 2012 Olympic sailing events in Weymouth, and the Lea Valley Olympic Park, London, for the 2012 Olympics.
- **Prestige project developments** (known as flagship projects) — such as waterfront developments (Cardiff Bay and the Welsh Assembly building) and convention centres (NEC, Birmingham). The major issue here is to what extent these projects directly benefit the people of the area. Also, much money can be wasted if projects are unsuccessful (e.g. the O2 arena for several years after 2000).

The problems

Environment
- Redeveloping extensive areas of derelict land, largely in the zone in transition around the CBD and in minor urban areas
- Cleaning up land contamination and water pollution which are barriers to development

Government
- Big issues — increasing disenchantment of electorate
- Disenfranchising, disinterest, detachment of urban poor

Socio-cultural
- Cities are increasingly divided, with polarisation of rich and poor
- Ghettos and areas of social exclusion are almost impossible to regenerate, even with a huge investment

Economic
- Massive decrease in the number of jobs because of deindustrialisation
- Centrifugal forces have encouraged industry into semi-rural greenfield sites
- Consequences of very high unemployment or under-employment in low-paid jobs

Image
- Many former industrial cities have a negative image of dereliction, dirt, danger from crime, decay and deprivation — not places to invest in

Figure 29 The major issues of regeneration

A number of governments have consistently tried to regenerate declining urban areas. The British case study below summarises the stages and the schemes. Such schemes reflect the political orientation of the government in power.

Phases of urban regeneration in the UK

Phase 1: post-war recovery — focus on housing 1947–67

The main aim was to provide people with housing. Many cities were heavily bombed.

- **Comprehensive Redevelopment Area (CDA)** programmes demolished slum areas and re-housed people in high-rise blocks (tower blocks). At the same time, utilitarian council estates were built in the outer suburbs (e.g. Leigh Park and Paulsgrove, Portsmouth). Some housing was erected as prefabs — emergency homes.

- **New Towns** — not inner-city but a part of the process that removed population from the inner city, e.g. Cwmbran and Crawley. Also, **Expanded Towns**, e.g. Andover, were designed for the overspill from urban areas in what were seen at the time as quality environments.

Despite these efforts housing shortages were still widespread and the dwellings were of poor quality. Redevelopment did not keep pace with decay and slum clearance led to vast areas of dereliction.

Phase 2: social and economic action 1967–77

Attention began to focus on the inner-city zones that had experienced **decentralisation** and **deindustrialisation**. The arrival of immigrants into, for example, Brixton (London), Chapeltown (Leeds) and Moss Side (Manchester), compounded some of the problems.

- In 1968, the **Urban Aid Programme** established the idea of government grants for a range of small-scale projects in deprived areas: housing renovation, advice centres and nursery schools, as well as community development projects.
- In 1969, **General Improvement Areas (GIAs)** were established and in 1974 **Housing Action Areas (HAA)**.
- **Education Priority Areas (EPAs)** were established in 1967 followed by **Community Action Programmes (CAPs)** in 1974 and then a host of other small initiatives.

There were many ideas, but little coherence and insufficient funding and therefore it was difficult to see lasting, tangible benefits.

Phase 3: Inner-city focus 1977–90

The government recognised the problems of inner-city areas and focused efforts on economic regeneration, environmental improvement and creating services for communities.

- **Partnership Areas** (e.g. Salford), which received annual central government funding for 23 programmes including the Govan area of Glasgow.
- **Urban Development Corporations (UDCs)** were introduced in 1981 as a property-led development to regenerate large tracts of land (e.g. London Docklands and Cardiff Bay). UDCs had similar powers to New Town authorities and were answerable to central government. Issues included: job creation, over-reliance on property development (which was only beneficial in times of rising prices), reduction in the power of local government, and public money used to encourage private development.
- **Enterprise Zones (EZs)** were also introduced in 1981. These covered small areas of land (under 450 hectares) and were expected to offer special incentives to attract high-tech industries by relaxing planning regulations. They were sited in areas of high unemployment, e.g. Gateshead and the lower Don Valley, Sheffield. Doubts were cast on the number of new jobs created because jobs migrated into the zones depleting the surrounding areas of jobs. Many zones failed to attract innovation.
- **Task forces** were small scale teams that acted as think tanks to saturate a deprived area with aid and grants (e.g. St Paul's, Bristol, and Moss Side, Manchester).

Phase 4: new ideas 1990 onwards

- This phase is associated with competitive bidding — the best projects get the money. **City Challenge 1991** involved private, public and voluntary bodies working together. The areas involved had major problems: high unemployment, low skills base, environmental dereliction and deteriorating housing (Wolverhampton) and housing improvements (Hulme, Manchester). While there were many fine projects with good private sector investment and imaginative ideas, the competitive situation also wasted energy and time.
- **Single Regeneration Budget (SRB)** was introduced in 1994 to join up all the sources of funding to make a coherent regeneration project. Cobridge, Stoke-on-Trent, (containing the red light district) was a good inner-city scheme.
- The variety of projects was bewildering. With the addition of millennium schemes like the Millennium Stadium in Cardiff and the O2 stadium at Greenwich, the public reaction was mixed.

Three trends emerge from this case study:

(1) The increasing role of central government in the inner city.

(2) Increased private investment usually following government 'pump priming' or 'matched funding'.

(3) Increased movement away from social schemes, with economic consensus seen as the key to improvement. Recently there has been a move back to education and social schemes.

When evaluating urban regeneration there are a number of criteria to use, particularly: were the objectives of any particular scheme met, and were the targets reached within the allocated period of funding?

Figure 30 sets out an evaluation framework. You should use this for your chosen urban regeneration scheme (e.g. Cardiff Bay, Swansea SA1). While there have been successes — cities such as Glasgow and Birmingham have transformed their images — a number of concerns have been expressed about regeneration in other cities, especially those carried out by the Urban Development Corporations (e.g. London Docklands).

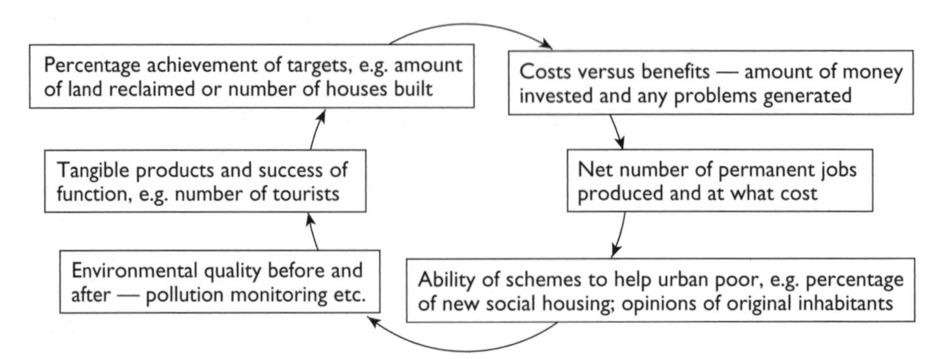

Figure 30 Evaluation framework for urban regeneration

Traffic and the inner city

Traffic congestion is a feature of everyday life in most cities. It threatens health and well-being. Congestion is a problem and there are costs associated with it that impact on the inner city:

- Air pollution — affects health of inhabitants, usually the most vulnerable.
- Noise pollution — constant traffic noise from routes in and out of CBD.
- Death and injury from accidents — again affecting the young and old.
- Costs in terms of well-being.
- Unsustainability of the energy used in commuting into and around the city.

Solutions to traffic problems in the inner city

- Build and redesign roads and traffic schemes — not very practicable in the inner city because it cuts off areas from one another.
- Promote public transport.
- Traffic management including traffic calming measures to stop rat runs through the inner-city streets.
- Discourage use of private cars and discourage no-charge parking areas in the inner city.
- Make cities more compact including the inner area.

2.4 What are the issues in the CBD?

The nature of the CBD

CBDs become more distinct the larger the city. A variety of regions can develop within the CBD. These regions are frequently a consequence of the changes taking place in society and the economy and the response of city governments to those changes. The key regions of the centre are:

- An old centre that was the medieval core of the city, which contains the cathedral or major church and old coaching inns, the old street market and some recreational functions such as restaurants.
- The modern retail core which may have developed out of the historic core. This is characterised by major national chain stores that have caused people to refer to High Streets as **clone towns** because the image presented to the public by each store is the same in every town. In the core of the retail region the function will cover all floors. It will also be the area of greatest pedestrian flows (**footfall**). In some cities there may be conserved aspects of former retail architecture, e.g. covered arcades as in Cardiff and Leeds. Often the area might contain a modern multi-storey centre, e.g. St David's in Cardiff, and the Arndale in Manchester.
- In some towns there is a smaller area of retail outlets, often specialised, that cannot afford the high rents of the inner core.

- **Quasi-retailing** is the name for banks, building societies and other financial outlets that in medium sized cities can often be found in a separate area.
- The office district has grown to be a separate part of the modern CBD. The larger the city population, the greater the scale of this main employment focus. In the largest cities the office area can have regions within it, for example, the legal profession may be in a distinct area, possibly near the courts. Separate stock-broking areas and insurance districts may develop, e.g. City of London.
- Some towns and cities have a governmental area where council offices, the police and law courts might be found. In capital cities this will form a large swathe of the central area such as Westminster, Whitehall and Victoria Street in London and the administrative offices clustered north of the centre in Cardiff.
- One of the growing features of the modern CBD is a recreational district of restaurants, bars and clubs. By day these areas are functioning to support the CBD but by night they can become no-go areas for many people.

CBDs spread in the direction of the growth of a city. Growth can be constrained by areas such as a Victorian park. Rivers also act as natural barriers to growth except in the largest of cities. Rail lines can also constrict growth.

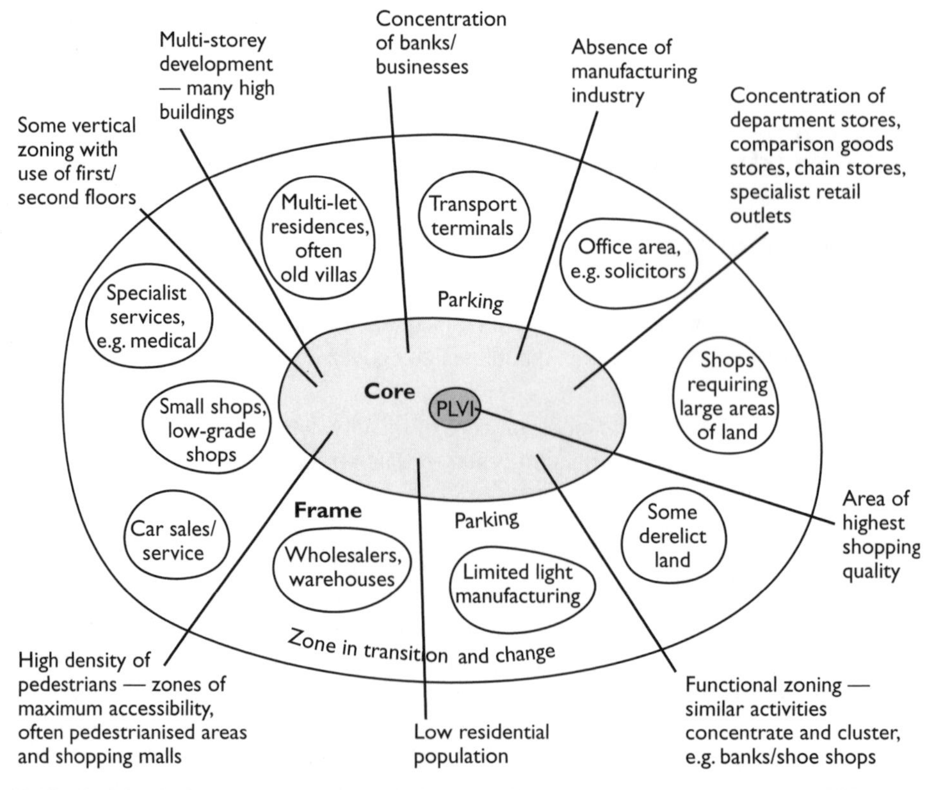

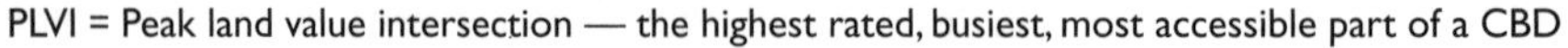
PLVI = Peak land value intersection — the highest rated, busiest, most accessible part of a CBD

***Figure 31* Key features of the CBD core and frame**

The changing CBD

Most CBDs in cities in MEDCs are currently under threat for five main reasons:

(1) The loss of retailing functions to out-of-town shopping centres and **retail parks**.

(2) The loss of offices to suburban or peripheral locations in prestige **science parks**.

(3) The nature of the CBD itself, which because of congestion is becoming increasingly inaccessible, especially as the bulk of the population lives in suburban areas. Equally, the CBD has very high rents and council taxes and is only affordable to prestige businesses, so the centrifugal forces begin to dominate.

(4) The effects of the 2008/09 credit crunch with the closure of retail chains, e.g. Woolworths in the UK, and the shrinking of the financial sector.

(5) The growth of central recreation areas.

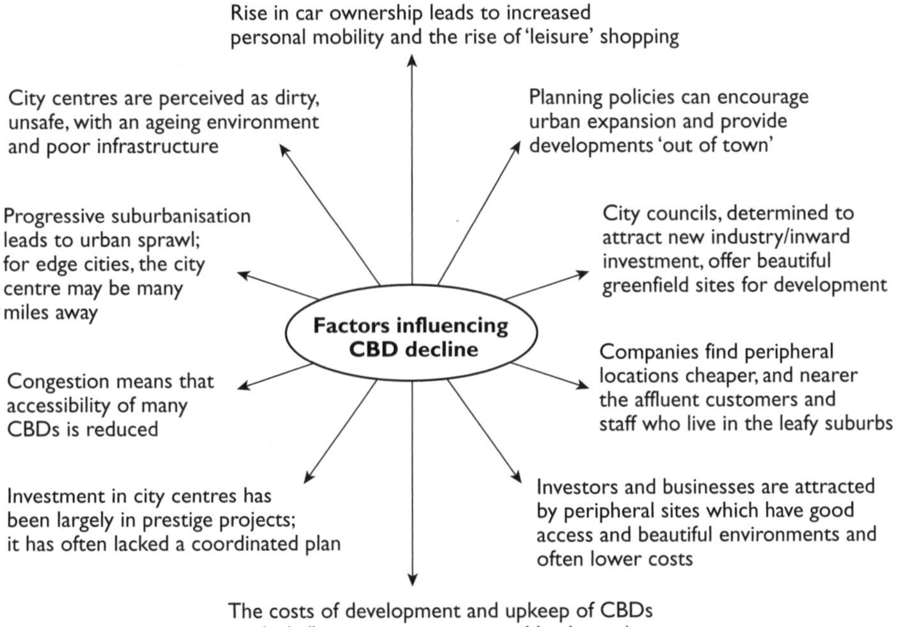

Figure 32 Factors influencing CBD decline

Reviving city centres

A number of strategies have been devised to enable city centres to remain vibrant:

- Using a city business and marketing team to coordinate the management of the city centre and encourage special events.
- Pedestrianisation to improve safety and to provide a more attractive shopping environment with new street furniture, floral displays and landscaping.
- Encouraging the construction of all-weather shopping malls/centres (air-conditioned in summer and heated in winter) in key locations in the city centre, often with integral parking.

- Encouraging the development of specialist areas, including attractive open street markets, cultural quarters and specialist arcades selling high-quality products such as organic foods and local handicrafts.
- Improving public transport links with rapid transit into the central, pedestrianised area, e.g. park-and-ride, shopper buses.
- Planning car parks to give ease of access to the centre whilst also controlling pricing and time parked.
- Making the centre safer with video surveillance. (This may be a perceived benefit as only 1 in 10 street crimes have been solved using video evidence.)
- Developing new leisure and cultural facilities.
- Promoting tourism activities to encourage greater spending by **conserving the heritage** — used in Bath, York, Chester and Cambridge to great effect.

In the USA, where shopping malls are widespread, the movement to reinstate and revitalise the town centre is strong as both planners and the general public realise the dangers of dead city hearts (**urban doughnut effect**).

Round-the-clock city centres

British city centres are frequently 'dead' and dangerous at night. By encouraging shoppers to stay longer and perhaps dine out, they may be encouraged to spend more. Here are some possible schemes:

- Run shopping events such as farmers' markets and Christmas Fairs.
- Late night shopping and Sunday opening.
- New retail developments focused around a major 'magnet' store, e.g. West Quay, Southampton, with John Lewis as the focus and IKEA nearby.
- Plan for a wider range of leisure facilities such as cafe bars, music venues, theatres and cinemas that people would naturally visit during an evening. Some of these will be located on the CBD fringe.
- Promote street activity by permitting cafes to spill onto the street (in the UK, only realistic in summer).
- Develop nightlife and clubbing — although the costs may exceed the benefits.
- Plan for theme areas such as the gay area of Canal Street in Manchester or the cultural areas of Stoke and Sheffield.
- Bring back residences into the dead heart — flats to rent above shops or the development of old buildings as flats (a strand of gentrification) targeted at higher-income workers.

Most city managers are trying a range of these strategies and shoppers are being attracted back to centres. These strategies can take place in conjunction with controlling permission for and limiting the number of suburban (e.g. Westfield Mall, Shepherds Bush), out-of-town shopping centres and giant super-malls such as Bluewater in Kent.

There are a number of studies on the impact of super-malls on traditional town centres. Usually the smaller towns are most affected. For example, Merry Hill in the West Midlands has affected Dudley far more than it has Birmingham.

Tackling Question 3c through CBD study

Question 3c in your examination will expect you to be able to write about **your own study of a topic in human geography**. CBDs are a good topic to select. You will be expected to gather information, know sources of information, have acquired the skills of mapping and field observation, interpret information realising that there may be bias, and draw conclusions from that information. CBD study provides the ideal opportunity for you to develop these skills through a field exercise in the CBD. Your approach to this part of your study of Changing Human Environments should follow a modified sequence of enquiry.

CBD enquiries that you could attempt

- An analysis of retailing in the CBD.
- The distribution of one retail type in a large CBD.
- Pedestrian flows and retail types.
- Street markets and pedestrian flows when a market is in operation and when not in operation.
- Differences in retailing between pedestrian streets and streets with traffic, or between arcades, streets and covered centres.
- Analysis of distribution of outlets in a new covered centre.
- The distribution of banks, building societies and other financial services.
- The changing nature of leisure in the CBD.
- Day and night leisure.
- The nature of the frame.
- Is there a zone of discard?
- Where are the empty shops?
- Is the centre a 'clone town'?

The sequence of enquiry

(a) Plan the enquiry

(1) Read about the CBD.

(2) Decide on a topic that is either an area of investigation (e.g. defining the edge of the CBD), or a question that you wish to answer (e.g. do shops cluster in the CBD?).

(3) Plan how you will get the information and data from primary (fieldwork/questionnaires) and secondary sources (Goad plans) that will enable you to answer your research topic. Work out the categories of shops that you wish to map. Prepare data collection sheets. Do you need others to help you?

(b) Carry out the fieldwork

(1) Record your data on data sheets.

(2) Know the faults or errors in your data collection.

(c) Process the information

(1) Map the distributions. Understand what the information that you have gathered actually says about the clustering of shop types that you are studying. How accurate is that information? Is it biased? If so, why?

(2) Be able to process data statistically (you will not be asked about the statistical processing in the examination but you might need to know what the statistical processing shows).

(d) Write 1000 words on your study

(1) Organise the information and data so that it answers your question. Which shops cluster — shoe shops? Do large stores cluster? Why do some cluster while others are spread out?

(2) Draw conclusions from your results that you will be able to use in the examination.

(e) Evaluate your work

(1) Evaluate each stage of your study so that you can answer questions about your sequence of enquiry in the examination.

(2) Evaluate your evidence and conclusions in relation to what you read at the outset and what you have found. This should include altering the topic or the question that you asked so that others could gain a better understanding of the topic in the future.

Questions that could be asked on your study

Both of these questions address stage (a) in the sequence of enquiry above:

(1) Outline how you planned your study into a changing human environment.

(2) Outline the planning stages of an investigation that you have undertaken into a changing human environment.

The following two examples address stage (b):

(1) Describe the methods that you used to gather data and information for an investigation that you have undertaken into changing human environments.

(2) Evaluate the methods of data collection that you used for an investigation that you have undertaken into changing human environments.

The following are two possible questions that address (c):

(1) Outline the methods that you used to process the information for an investigation that you have undertaken into changing human environments.

(2) To what extent and why was your information biased that you gathered for an investigation that you have undertaken into changing human environments?

Two possible examples of questions that address (d):

(1) For an investigation that you have undertaken into changing human environments, outline the conclusions that you made to your study.

(2) Discuss the main findings of an investigation that you have undertaken into changing human environments.

Finally, questions on stage (e) might ask:

(1) To what extent is it possible to increase the validity of the conclusion of the investigation that you have undertaken into changing human environments?

(2) Attempt an evaluation of your sequence of enquiry for an investigation that you have undertaken into changing human environments.

All these questions could be answered if you carried out an investigation on the CBD.

2.5 How is the rural–urban fringe changing and why?

The rural–urban fringe emphasises the concept of the rural–urban continuum (see Figure 33). It is very much a zone in transition — a highly desirable area for which many land uses compete. Urban sprawl gradually extends into the rural–urban fringe and causes many pressures and conflicts.

(1) The **urban fringe** is a zone of advancing suburbanisation — with high levels of car ownership. High-quality houses on the fringe of an urban area are always in demand — the urban fringe is perceived as quieter, less polluted, more crime-free and less congested, often with access to quality services such as good schools. With perceived beautiful views over countryside, living at the urban edge is seen as combining the advantages of town and country living. In general, the homes built here are more highly priced and sell more easily. But as the urban fringe is always advancing (unless it is controlled by a green belt), the rural 'idyll' may become, after a few years, an illusion as the open view disappears and land is gradually released for development or ring roads.

(2) The **green belt** is used in European countries to control urban sprawl. It is an area of open farmland and recreation land around an urban area (mineral extraction is permitted). The only development that can take place in green belt areas is within existing settlements (**infill**), or where planners decide that local, regional and national needs will be met by an extension to the settlement. The Cambridge Science Park was a case in point because it brought new jobs to the region. In theory, by restricting growth in the green belt, developers are being encouraged to use **brownfield** sites in inner urban areas. The reality is that because of the higher cost of developing brownfield sites, the developers merely start building beyond the green belt in the commuter belt, a process sometimes called **leapfrogging**. There is evidence that green belts are being reduced in size, but they are also very restrictive. Some planning authorities form **green wedges** or **green corridors**, e.g. between the Ruhr towns in Germany, because they allow more development yet also conserve open countryside. Because of their rarity value, any developments in the green belt can command very high prices and housing that pre-dates the creation of a green belt soon becomes exclusive.

(3) The **commuter belt** has developed beyond the green belt in response to demand. Where there is a motorway or fast road, large dormitory settlements develop — so called because commuters only come home to sleep there. Technically, as these occur in semi-rural areas, they are the products of the process of **counter-urbanisation**. But these settlements are very suburban with large estates of private housing, out-of-town supermarkets and leisure facilities. However, in the sectors between the main transport routes, commuter developments are more restricted — often to small estates of under 10 houses, lifestyle hobby farms or renovated country cottages. Here the process is genuine counter-urbanisation. Counter-urbanisation is also the product of firms moving to small towns and taking their workforce with them. Some people also move to this zone for retirement but are dependent on housing where there are basic village services.

(4) Beyond these zones are the beginnings of a rural zone that is used by city dwellers for weekend recreation. Clearly the areas under most pressure are the attractive and accessible areas. **Country parks** have been developed in this zone specifically for day trippers. The original idea of these country parks was to take pressure off high value, environmentally sensitive sites in National Parks.

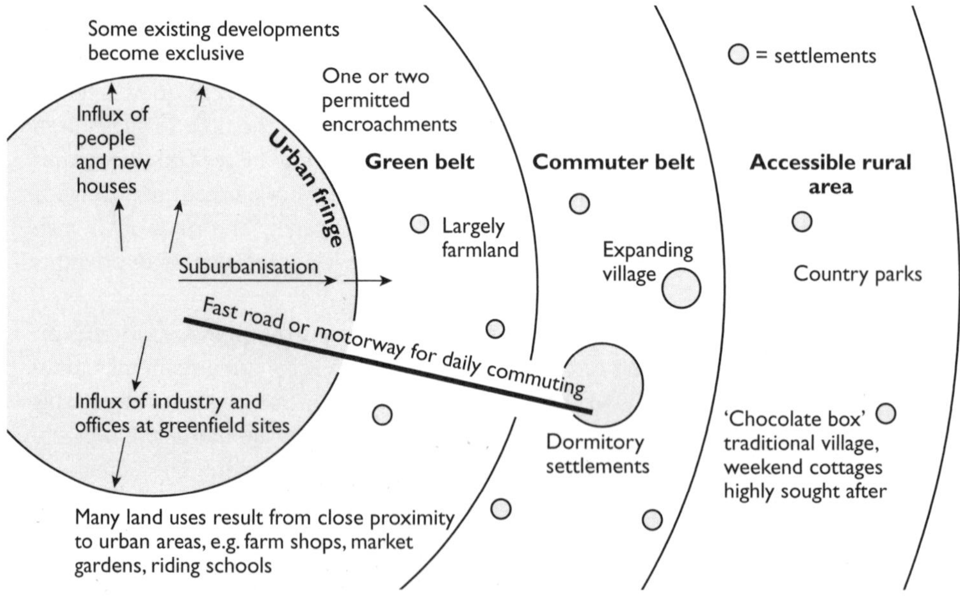

Figure 33 Main features of the rural–urban fringe

Types of development in the fringe

- Intensive and/or agribusinesses such as market gardens and pick your own.
- Playing fields used by urban dwellers for formal and informal recreation.
- Golf courses, shooting ranges — extensive recreation.
- Park and ride terminuses (e.g. Oxford).
- Science and Technology Parks (e.g. Cribbs Causeway, Bristol).
- Hotels like Travelodge and Holiday Inn Express.

- Major sports stadia (e.g. Falmer Stadium, Brighton).
- Industrial sites (e.g. Rolls Royce in a disused gravel pit near Chichester).
- Retail and office parks (e.g. Hedge End, Southampton).

The mixed nature of the rural–urban fringe is influenced by four main factors: agricultural policies; the nature of the agricultural economy; the urban economy; countryside planning regulations.

The rural–urban fringe is a mosaic of land use and land types — from high quality to degraded or derelict land — all within a small area. Many land uses look attractive but others can be ugly. Because the area is under pressure from competing uses, its state and appearance can change rapidly. Also, many land uses, like golf courses and sailing in former gravel pits, are not very natural and are controversial. Conflicts result from:

- **Commercial decentralisation** — industry seeking rural greenfield sites as opposed to city brownfield sites.
- **Housing demands** place pressure on the rural fringe — a strong desire to build in green belt areas and attractive villages.
- **Farming** — problems of trespass, vandalism, fly tipping and dogs, where people seeking recreation try to use farmland.
- **Environmental issues** — much of the development of roads, new science parks and new housing can cause increases in pollution and also heighten flood risks.

Threats to and issues in the rural–urban fringe that may form a worthwhile investigation:

- Retail parks and superstores and their impact on the environment and other retail outlets.
- Leisure centres and new parks on reclaimed land such as former gravel pits.
- Marinas in reclaimed areas such as Maryport and Port Solent.
- Loss of school playing fields.
- Office and science parks and impact on commuting and employment.
- Problems of intrusions onto rural land — fly tipping, horticulture.
- Former airfields now engulfed by urban areas (e.g. Brooklands, Surrey).
- The impact of leisure developments such as golf courses, paint balling, mountain biking, motocross, leisure centres and rowing/sailing lakes.
- Bypasses — these can be studied but don't make very good investigations.

2.6 How are rural settlements changing?

How and why do rural environments vary in character?

There are an enormous variety of rural environments and therefore settlements within them. MEDCs are becoming more urbanised in that they house people who are urban in their way of life. In most MEDCs people are psychologically urbanised because they expect the same services in rural areas, e.g. broadband, cable television etc. The factors affecting the nature of a rural settlement can be the product of: distance from an urban area; location in the country; whether the area is conserved by legislation, e.g. in a National Park; land holdings in the area; the architecture of the settlement.

The nature of rural settlements

A number of terms are used to describe the nature of the settlements that form an important part of the rural landscape. You will be familiar with most of these.

- **Distribution** — how farms, villages and dwellings are spread across the landscape.
- **Density** — the number of people per km^2. This will have a major impact on the number and concentration of dwellings, farms and villages.
- **Hierarchy** — a functional system based on the importance of each settlement in an area. Importance is a function of population size and range of services. The higher up the hierarchy, the lower the number of settlements and the further apart they are.
- **Dispersed** — settlement is in the form of scattered, isolated farms and dwellings.
- **Nucleated** — dwellings are concentrated together to form hamlets and villages in favoured sites, for example near a spring or at the junction of mountain valleys. The shape of the nucleated village can be linear or compact.
- **Site** — the nature of the land on which the settlement is built. Favourable sites are often flood free, south facing, sheltered, on a spring line, etc.
- **Situation** — the location of a settlement in relation to its surrounding area.
- **Function** — the purpose of a village, e.g. agricultural, leisure, mining.

There are certain characteristics that are common in most rural areas. It is possible to develop an **index of rurality (ruralness)** based on whether an area shows these characteristics or not. In 1971, Cloke identified 10 key indicators of rural life, devised from the census for MEDCs. Table 8 lists the indicators that apply to rural settlements and assesses how useful Cloke's indicators are today.

Table 8 Cloke's Index 1971 (revised)

Revised 1971 indices	Characteristic 2009	Change since Cloke wrote index in 1971
% of workers in agriculture and forestry	Very low in villages	Declined since 1971
% of population over 65 years old	More retired in rural areas	Distinguish rural retired from immigrant retired
% of working age population 16–65	Varies with distance from employment centres and proportion in higher education	Counter-urbanisation brought more workers into rural areas
% population change among resident population	Rising due to in-migration but falling in remote rural areas	Second home population has risen — more in remote areas
Distance from urban area	Greater the distance, the more rural	Still true
% employed outside of settlement	High in commuter and counter-urbanised areas and lower further away	Teleworking, working from home one day a week, are altering this
% resident less than 5 years	Measures recent immigrants	Slight increase but misses second home owners

Revised 1971 indices	Characteristic 2009	Change since Cloke wrote index in 1971
Population density	Declines with distance although rising with infilling. Lower in remote areas	Still same as 1971
Occupancy rate: % of population at 1.5 persons per room	A general trend not just in rural areas that shows decline	Measures little in 2009. Time distance to service provision would indicate more
Household amenities (baths, central heating)	Better measures would be car ownership and access to mobile phone network	A dated measure in twenty-first century

Cloke's criteria did identify a spectrum of rural settlements but, as the right-hand column shows, many of these criteria are inappropriate almost 40 years later. There have been great changes in rural areas, so much so that alternative criteria might include:

- Ability to receive signal for mobile phone.
- Access to a supermarket, secondary school, and health care.
- Percentage of built-up land in parish.
- Access to mains gas.
- Percentage owning two cars (now a necessity for many rural families).

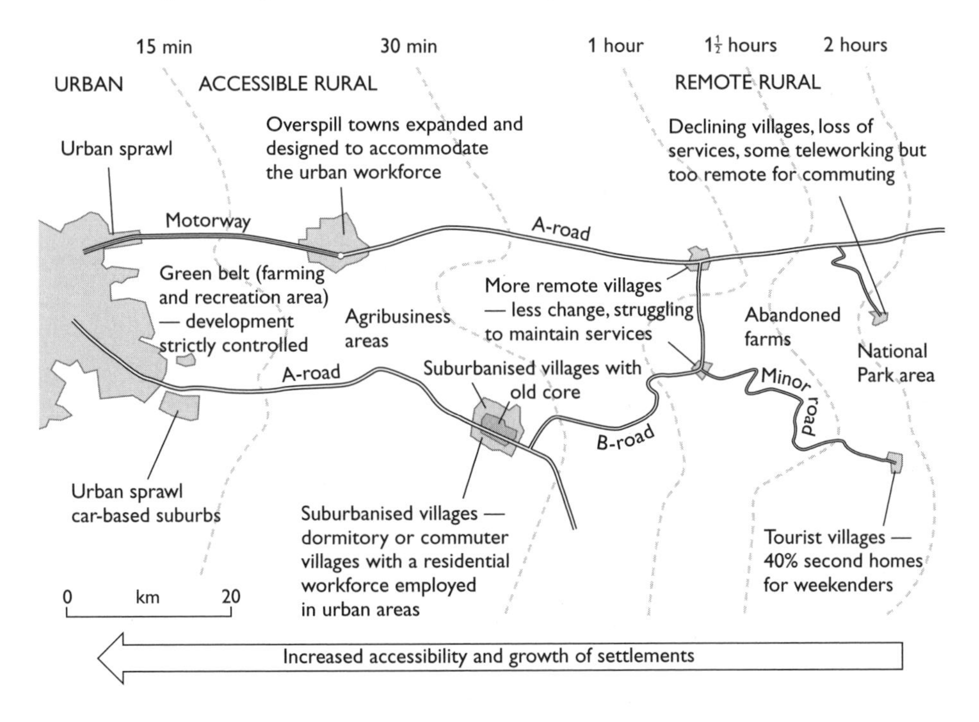

Figure 34 The range of rural settlements with distance from a city

Changes in rural population and settlements

Although, in general, remote rural areas lost population in the period 1930–1970, in most MEDCs, accessible rural areas gained population — essentially as a result of the movement of people and employment from urban areas. The movement was facilitated by improved public transport and increased ownership of cars. From 1970, as **counter-urbanisation** began to take place, most rural districts began to experience an upsurge in population known as **rural turnaround.** Rural areas (even the more remote rural areas) increased their population by an average of 16% between 1971 and 1991. This is in contrast to metropolitan cities where populations decreased by 18% in the same period.

Counter-urbanisation

The factors that led to counter-urbanisation vary from settlement to settlement but the following are among the most common explanations:

- The planning process that created new towns, expanded towns and permitted small towns to expand, led to more jobs within easy reach of rural settlements.
- Improved road communications (motorways) and faster rail transport.
- Improvements in communication (e.g. teleworking using the internet).
- Growth of branch factories (branch plants) in the new planned settlements.
- People will take jobs in scenic areas, sometimes for lower salaries and will trade off salary for a better environment.
- Green belt legislation that made areas more desirable.
- Diseconomies of scale in cities.
- National shift towards service employment and specialised rural manufacturing growth.

The consequences for rural settlements

- Reduced availability of homes for those raised in the countryside and for rural workers.
- Some villages have been planned to take social housing and lower paid workers, e.g. Lavant, near Chichester.
- A new service-sector class works in the villages, e.g. cleaners, small builders.
- General stores close because people shop in the superstores in cities and have cars that enable them to shop away from the village.
- Counter movement of some towns and large villages attempting to retain local shops, and banning the large chains from opening stores (known as Tescoisation).
- Closure of post offices because business has declined (e.g. pensions no longer collected from a post office).
- Services for the more affluent such as antique shops — Arundel in West Sussex has one general store but 10 antique shops and art galleries.
- Villages in the most desirable locations in green belts attract the affluent, many of whose children have already left the home.

Figure 35 and Table 9 summarise the possible effects of change on declining and expanding villages.

Strong retirement element or people who have lived there for years

Fewer old people

New arrivals who commute to work (91% have car access or work from home)

Waisted shape — young people move away to look for work or cheaper housing

Expanding population — a baby boom from young families who have moved into area

Regressive base — few children born

90+, 85–89, 80–84, 75–79, 70–74, 65–69, 60–64, 55–59, 50–54, 45–49, 40–44, 35–39, 30–34, 25–29, 20–24, 15–19, 10–14, 5–9, 0–4

15% 10% 5% 0% 5% 10% 15%

Declining population structure	**Expanding population structure**
The village is declining because of out-migration, and the loss of the productive element has led to a further decrease. It is a very ageing pyramid	The village has expanded because of new in-migration which, because of the age profile, is providing further natural increase
Consequences of decline	**Consequences of expansion**
• As villages decline, many of the people left behind are old and of limited means • As people move away because of a lack of access to employment or affordable housing, other services, such as post offices, village schools and buses, become uneconomic, so even more people move away • Houses are bought for second homes, creating a ghost town for much of the year • Deprivation sets in, so many of the people left cannot move away and lead very restricted lives	• Pressure on key services such as school and health • Creation of several new 'small' housing estates, often executive houses local people cannot afford (could be called rural sprawl) • Little impact on village services, as newcomers rarely use village shops or garages — but they often use the pubs • Increased traffic congestion, especially at peak times — many new families have two or even three cars • Many new villagers do contribute to aspects of village life such as parish councils, Women's Institute (WI) and the church, but most do not • Villages are often dormitory villages with little traffic during the daytime • Conflicts can occur between traditional villagers and newcomers — traditional villagers feel their values are not respected
Breaking the spiral of rural decline and deprivation is the key issue	The main problem here is trying to maintain the rural community and a rural quality in the face of increasing urbanisation

Figure 35 Rural decline and expansion — the geographical consequences

Table 9 Why have rural services declined and why might they improve?

	The bad news	The good news
Food shops	• Supermarkets opening even in quite small towns, with lower prices, extended hours and free buses from some villages	• Relief from uniform business rates • Development of new types of village shops, e.g. farm and garage shops
Post offices	• Small offices downgraded to part-time community office status – unattractive as going concerns for people to run • Diversion of pensions business to banks	• New deals for rural post offices with banks, with the post office counters for general new business
Public transport	• Bus deregulation led to closure of uneconomic routes • New unitary councils subsidise their own services, not rural cross-council links • Not enough passengers	• Grants for community taxis and buses which are often developed by parish councils
Village schools	• Ageing population leads to a lack of 'customers' • As schools compete for numbers the tendency is for children not to attend their local village school, or they go 'private'	• Development of nurseries • Generally good performance of small schools encourages numbers • Clustering strategies for a shared headteacher • Government grants for small schools
Libraries	• Costs of provision • Cuts in council services	• Increased mobile libraries
Primary healthcare	• Closure of GP 'branch' surgeries • Decline in NHS dentistry • Essentially escalating NHS costs	• Creation of mini health centres • Extra grants for rural GPs • Development of rural pharmacies • Use of internet for consultations
Village halls	• General trends of changing family habits • Withdrawal of funding for youth clubs and social services	• Millennium grants for refurbishment

Rural poverty and deprivation

Rural poverty is a problem in MEDCs. Rural areas in MEDCs are largely regarded as more affluent than urban areas, but they contain a significant proportion of poorer people, scattered thinly, with small pockets of severe deprivation. Because of the smallness of these groups, and the perception of rural areas as idylls, they can be easily hidden. **Rural deprivation** is usually associated with a lack of material resources and is really a 'state of non-well-being', including impacts on physical and mental health caused by this lack.

Some peripheral areas, such as rural Cornwall, are now eligible for Category 1 EU development funding (the most extreme hardship) because these areas have some of the lowest average earnings of anywhere in the UK. Equally, while unemployment is generally lower in rural areas, finding employment can be difficult for school leavers. There is often a limited range of job opportunities because of physical constraints, and much employment (around 27%) is part-time and, therefore, lower paid.

Rural areas contain a greater proportion of elderly people (20%) and while housing in rural areas is generally of a higher standard, there is little affordable housing in villages. The poverty of these mainly elderly rural folk can be extreme, especially when it is combined with deprivation of accessible basic services, and the fact that people are less willing to declare their needs. Rural **social exclusion** exists, where people are excluded from social, economic or cultural opportunities because of low income, poverty, poor health or reduced access to services. **Social exclusion** is the process whereby the various systems that should guarantee the social integration of individuals or households fail to do so, possibly due in part to failure of employment provision or failure of housing markets.

Rural deprivation takes three forms:

(1) **Household deprivation** is concerned with the hardship of individual households trying to maintain a living standard. Examples are the plight of tenant farmers in less favoured areas, such as the Welsh mountains, or that of elderly single people solely dependent on their state pensions. Household deprivation is typified by poor-quality housing and high levels of income support and benefit payments.

(2) **Opportunity deprivation** is concerned with country dwellers' lack of access to education, health, work, social services and shops. Many rural families have to face the expense and difficulty of travelling longer distances for basic services, such as fuel, as well as paying more for them.

(3) **Mobility deprivation** is a measure of the lack of transport for going to work or for obtaining basic services, many of which are now concentrated in key settlements because of rural rationalisation of settlements. Owning a car or a motorbike is extremely expensive, especially for low income families. Access to a village taxi, post-bus or community bus service can be vital in the absence of commercial bus services.

The combined impact of the various facets of deprivation can, unless arrested, cause a cycle of **rural decline.** This can lead to rural **depopulation** if the people lost by out-migration are not replaced by newcomers.

Questions & Answers

There are three questions in the exam based on the content of Unit G2 (see page 4):

The levels marks scheme used for parts (b) and (c) of questions 1 and 2 and part (c) of question 3

Level	Marks	Descriptor
3	8–10	Very good knowledge and understanding with a good use of examples. Full descriptions and explanations. There is both breadth and depth to an answer that may draw together a variety of points. Follows the commands in the question. Has a good mini essay style, well written with a good use of the language of geography.
2	4–7	Some knowledge but not always clear that information/principles are fully understood. Information and explanations are partial. Sound descriptions. There should be more detail. Essay style present but probably lacks paragraphing and introduction/conclusion. Some use of the language of geography.
1	0–3	Limited and superficial knowledge with limited understanding. Little use of examples and if they are present it will be e.g. Africa (often quoted as a country when it is a continent). The material will be simple and probably a very brief one paragraph response with no development. Slips of grammar and spelling evident.

Command Words are the words telling you what to do with the subject matter. Do remember that there are commands or guidance on the cover of the examination paper. Read these as they stress use of examples, and the necessity for good English.

- **Describe** expects you to show what you know is occurring or what a map or diagram shows. It expects you to know what a process does, what happens, where it occurs, when it occurs. Who or what causes a process and whom it may affect can be part of descriptions.
- **Explain** expects you to say why and how something occurs. It might involve some description but do not rely on description as a means of explanation.
- **With the aid of a diagram** means that you must draw one. If it is a diagram it must be labelled and the axes on a graph must also be labelled.
- **Identify** will require you to provide listed points which are supported by examples.
- **Examine** expects you to provide reasons and to discuss both sides of an issue: for instance, 'Examine the costs and benefits of regeneration schemes in cities.'
- **Justify** expects you to give a reason.
- **Outline** is asking you to state the main points or factors and is expecting more than two points with supporting examples.

Examiner's comments

In this section each candidate's answer is followed by examiner's comments, indicated by the *e* icon. These comments show how the marks have been awarded and highlight specific problems and weaknesses and areas for improvement.

Theme 1

Q1: Fertility and death rates

Key

total fertility rate	long-term impact
less than 2.3	declining population
2.4–2.9	stable population
3.0–3.9	growing population
4.0 or more	rapidly growing population

***Figure 1** Total fertility rate by country, 2000*

(a) Describe the distribution of total fertility shown in Figure 1. (5 marks)

(b) Give reasons why countries experienced a total fertility rate less than 2.3, the replacement level. (10 marks)

(c) With the help of a diagram, describe and explain the changes in the death rate shown by the demographic transition model. (10 marks)

C-grade answer

(a) The African continent has the most rapid growth. Countries which have a total fertility rate of 2.4–2.9 are mostly MEDCs. These countries have a stable population as their standard of living is higher. Declining population is also found in MEDCs.

e This is a basic answer that only identifies one area. It accounts for one category, which was not asked for in the question. Therefore, it lacks any depth of description and would only gain 2 marks.

A-grade answer

(a) Rapidly growing population can be seen in much of Africa, Saudi Arabia and Pakistan and parts of Latin America. The population is growing less rapidly in India and Central America. The areas of stable population are North America, mid Eurasia. The countries with declining population are in Europe, North America and Australia.

e This answer covers all of the categories and provides examples of each category. It will gain 5 marks despite the generality of the description.

C-grade answer

(b) Many countries experienced a total fertility rate below the replacement level (less than 2.3) because in those countries the government is placing policies where families are not to have a large number of children. The countries where total fertility rate is below the replacement level of 2.3 are countries which are developed, such as in Europe, in Australia and some places in South America. The government places different types of policies, for example, in China, there is a 'one child policy'. This was placed because the population was very large and growing every year, but now they are controlling the population. The 'one child' policy is that if a family has not more than two children, they get a 5–10% incentive in their salary, but if a family has more than two children, they get a 10% reduction in their salary. Also in some countries people decide not to have any children because they want to work and have money to spend on themselves. They use contraceptives to stop having children.

e The answer provides two reasons although too much time is spent on the China policy rather than providing more evidence of other policies. The language does include repetition and only develops the points very generally. It does not refer to replacement level. There are no paragraphs in this mini essay, which is consequently rather short. It is a Level 2 answer and would gain 5 marks for its content and reasoning.

A-grade answer

(b) Most of the countries experiencing a total fertility rate of less than 2.3 are the MEDCs. Reasons for these countries having a total fertility rate below the replacement level are because there are higher percentages of women working in the work force in either fulltime or part time jobs. With their involvement in the workforce, women tend to want fewer children and at a later age as they want to concentrate on their jobs. In MEDCs, the healthcare system and pension schemes are very good, therefore decreasing the number of births as most children will make it to adulthood and parents would not need children to look after them when they are elderly as they will have a pension scheme. Contraception and birth control measures are readily available, therefore reducing the amount of unwanted pregnancies. Family planning schemes are also available, therefore enabling families to plan the number of children they want to have. Many people stay on in further and higher education, which will lead to them marrying at a later age and having fewer children as they want to concentrate on their careers first.

Sometimes there are government policies to stop or encourage people to have smaller families.

This mini essay is about the right length for 10 marks taking about one side of an answer book. It would benefit from better paragraphing and some words of introduction and conclusion. On the other hand it covers a range of factors which are well explained in a theoretical sense. Other factors that could be used are: societal norms for size of family, lifestyle preferences, costs of children, delayed parenthood and increased infertility, and singletons. It is difficult to provide examples of these factors although naming countries where this is the case would help. It is a Level 3 response and would gain at least 8 marks and be on the way to an A grade.

C-grade answer

(c) ***Demographic transition model***

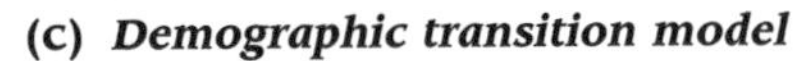

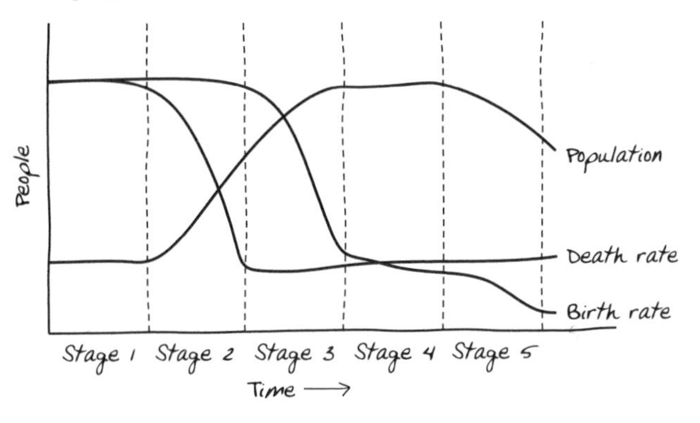

The DTM shows that in stage 2 the death rate drops causing the population to rise. The DTM shows the process countries go through in terms of economic development (the different stages) and its effects on population. As countries develop they can afford better technology and food. This means there is better hygiene and healthcare, so death rates drop until most deaths are caused by old age or accidents. As the DTM shows, this affects the population bringing death rates below birth rates and causing population growth. Another factor causing the low death rates in MEDCs which are in stage 4 of the DTM is because of improved medical facilities, diet is balanced and due to that water and food is not contaminated. Hence diseases such as cholera and typhoid can be avoided. In stage 3 there is a steady death rate. This means that the country will have a steady population growth. This is common in MEDCs such as the USA.

The candidate has not assembled the response into a coherent answer. The diagram is not accurate, especially for Stage 3. There is no mention of Stages 1 or 5 despite

them being on the diagram. Stages 2–4 are dealt with in a random order but the reasoning is on the right lines for Stage 2. Stage 3 is partly correct but the example is incorrect unless it is given a date such as 'late nineteenth century USA'. The answer lacks a description of the death rate line but does get credit for some of the explanation. It is a Level 2 response gaining 4 or 5 marks.

Overall, this candidate will gain 12 or 13 marks out of 25 for this response, which is very much in the middle of the marks range and indicative of a low C.

A-grade answer

(c)

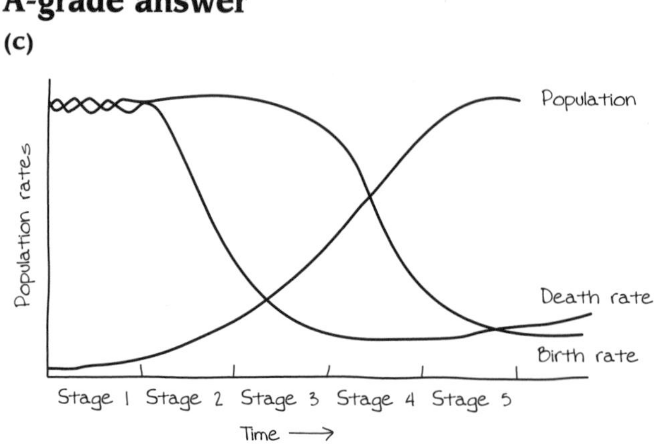

In the demographic transition model that I have drawn it shows that death rate has changed over time and many countries fit into the graph in each of the stages.

In the graph the death rate fluctuates very highly with the birth rate in Stage 1. This tends to occur during a war like during the Napoleonic wars and earlier. Countries in Africa such as Rwanda and Congo fit into stage 1 because of their high death rates caused by war, AIDS and starvation and a lack of access to clean water. In Stage 2 the death rate drops. This is from an improvement in medical care which saves lives and stops deaths. In stage 3 the drop in death rate slows down but it is still slowly dropping, this is because of the last of the older generations from when the death rate rapidly declined in stage 2, but the birth rate remains high. This was when people started to know how to stop disease and help keep people alive so that they lived longer. Stage 4 shows the death rate staying constant with a bit of a rise, this is caused by a famine or disease like when flu hit the population at the end of the first world war. Stage 5 the death rate rises above the birth rate to show decreasing population, this is just a guess as stage 5 has not properly occurred yet. Many countries fit into different stages like now Australia, UK, Sweeden etc. Italy and Japan also fit into stage 5 because they have ageing populations

e The candidate attempted a brief introduction but no conclusion that summed up the various explanations. The description is reasonably accurate. Explanations are present although the quality of expression can give the marker the idea that he/she doesn't understand. The examples are all reasonably accurate. The candidate's quality of English is variable and the fact that Sweden is misspelt at the end will be ignored by the marker although it will influence the quality-of-language aspect of the marks. There is enough description plus the diagram, which, when combined with the explanations, will just get this candidate a grade A despite the slip in spelling Sweden. It is a Level 3 response worth 8 marks.

Overall, this candidate would gain a possible 21 marks for the question, which would be an A grade.

Q2: Population growth, migration and mortality rates

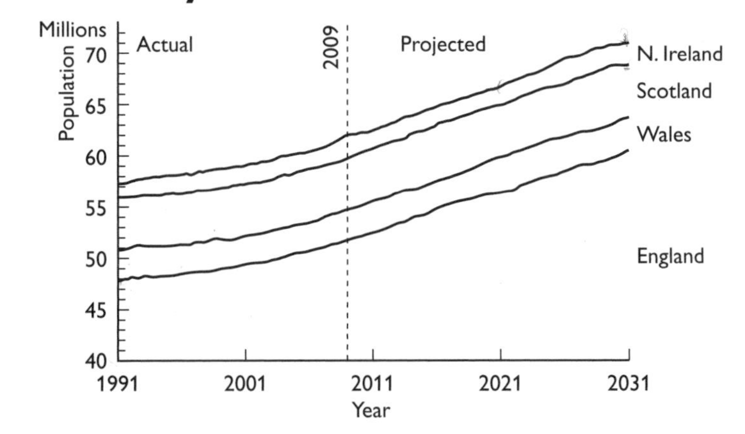

Figure 2 UK population 1991–2031 (source: www.statistics.gov.uk)

(a) Describe the patterns of population growth shown in Figure 2. (5 marks)

(b) Outline the consequences of the inflow of refugees and asylum seekers into countries with developed economies. (10 marks)

(c) Explain reasons for variations in mortality rates between countries at TWO different stages in the demographic transition. (10 marks)

C-grade answer

(a) Figure 2 shows that the population is growing and is projected to grow in England, Wales, Scotland and Northern Ireland. England's population has risen from nearly 48 million in 1991 to around 52 million in 2009. It is projected to rise to over 60 million in 2031. The total population of the UK has risen from over 57 million to around 60 million in 2009. It is projected to rise to 71 million by 2031.

There is only a description of two of the most obvious aspects of the diagram. The fact that it is a cumulative total is not noted and no comment is made about the other totals. Therefore, this response will only gain 2 marks.

A-grade answer

(a) Figure 2 shows patterns of population growth from 1991–2031. It shows that each of the countries in the UK has experienced a rise in population from 1991 to the present and all are expected to rise even further by 2031. In England the population in 1991 was 48 million. This has risen now to 51 million and is expected to rise further to 61 million by 2031. Overall in the UK the population was 57 million in 1991. It has risen to 62 million now but by 2031, is predicted to reach 71 million people. It is also interesting to note how the growth rate will increase in this time also. In the ten years from 1991 to 2001, the overall population of the UK rose by 2 million, however in the ten years from 2021 to 2031 the population is projected to increase by 5 million people.

e A range of valid points are made about absolute and English totals. In addition, the rise in the numbers was noted and described with support. In the time allowed this was a very good response. This answer will probably gain 5 marks.

C-grade answer

(b) In Caia Park in Wrexham, although the majority of people are from a white-Welsh background there is a small number of Iraqi Kurd refugees living in part of the estate. One Iraqi Kurd was subject to an unprovoked attack by 15 men with iron bars and knives. The next evening there was a riot which started in the Red Dragon Pub in which more than 200 people were involved. Petrol bombs were thrown and cars were set alight. These events highlight the conflict that is often created when refugees and asylum seekers move to a country such as the UK with a developed economy.

Other problems created are languages. Services such as schools and doctors may have to pay more money for translators to cope with the refugees and asylum seekers. This money often comes from taxes and could cause tax rates to increase.

Refugees and asylum seekers may also not be able to get any work in countries such as the UK because of the language barrier and many of the refugees and asylum seekers' qualifications may not be accepted. This gives them a big disadvantage when looking for a job.

e The example is good but it is about the flow into one area and the consequences are implied to be rioting. The example could be used more constructively to make a point. There is no attempt to define the terms. We are not told what the conflict is about. The answer is a mixture of vagueness and detail. There is no evidence provided in the last two paragraphs although the points may be valid. There is no reference to any gains for a country. The candidate gives the impression that he/she has not read much geographical literature on the topic and is relying on the media with all of its biases. There is no conclusion to the essay, which may struggle to get half marks.

A-grade answer

(b) There are many consequences due to the inflow of refugees and asylum seekers into countries with developed economies. In the UK, a total of 800,000 Eastern Europeans have entered the UK, with 500,000 of these from Poland. Although many of these are economic migrants, there are some refugees and asylum seekers from the former Yugoslavia in the figures. They bring with them many consequences: quite often, they don't speak the language of the country they are moving to. This has been a problem at Beeches Primary School in Peterborough, where there are 700 students, 25 different nationalities. Many translators are needed, costing the school and indeed the government money. Also, in some areas of the UK where there are not enough jobs, they can be seen as taking people's jobs. Quite often, though, they are the jobs that local people don't want to do such as flower picking.

They are cheap to employ, and are willing to work longer hours than local people. A further positive consequence is cultural benefits they may bring with them. These could be in the form of different religions, to fast food takeaways. Finally, the inflow may cause racial tension in many areas, and could form cultural sub-urbs and ghettos which would be bad for the local community. Another example of an inflow of refugees and asylum seekers into developed economies would be West Africans that travel through Libya (intervening place) to reach Europe where similar consequences are encountered.

e **Again, there are no definitions. The final sentence does suggest that the student knows what a refugee might be although early on the answer is more about issues of immigration in general. These could be made more relevant if the answer focused on Croats and Serbs rather than Poles. The consequences have the correct principles but not the best support in terms of the groups that are discussed. Nevertheless, the principles will enable this answer to gain 6 or 7 marks. If better organised, it would certainly gain 7 marks.**

C-grade answer

(c) In Italy, mortality rates are very low because there is a lot better healthcare than in a country such as Tanzania, which is at a lower stage of the demographic transition model. A free public health service was set up in 1978 in Italy. This allowed people who up until then could not afford good healthcare to get it. This increased life expectancy and lowered mortality rates.

Another reason that mortality rates are low is that since 1945 Italy has had a lot of industrialisation and urbanisation. This means that the economy of Italy has grown considerably and therefore more people are richer. Generally this means that the mortality rate is lower as people are able to afford a high standard of living.

In Tanzania there have been many problems with water-borne diseases such as cholera because of the lack of clean water in many areas of the country. This has increased the mortality rate significantly and has been one factor why the life expectancy is just 47 years.

Tanzania is generally a much poorer country than Italy. Therefore, many people are too far away from medical facilities and cannot afford to travel there or pay for the treatments. This leads to many people dying when if they had lived in Italy they may have survived.

The introduction names two countries but does not say at which stage each country is located. Two reasons for low mortality are stated but the candidate doesn't say which stage of the model the answer is referring to. Do not assume that the examiner knows.

Tanzania would be a good contrast if we knew the stage of the transition that was being discussed. There is only one figure quoted in the whole answer, which should have contained figures illustrating the different death rates. The answer is too general (even the candidate used the word 'generally') although the explanations have some validity. This response may get about half marks for the generalisations.

Overall, this candidate would probably score 12 out of 25 marks, which would be at the C/D border. Precise data, a knowledge of the DTM and better explanations would raise this mark.

A-grade answer

(c) Namibia and the UK are at 2 different stages of the demographic transition model. Namibia is at stage 2 of the model meaning that their mortality rate is very high, 20/000. This is due to a number of reasons. Namibia is an LEDC, meaning that their economy isn't as developed as those such as the UK. Because of this, there isn't as much healthcare, with medicines not always readily available, and this affects the mortality rate. Also education isn't as successful as in MEDCs, so people aren't always educated or made aware of sex and disease. The lack of contraceptives means that AIDS is a common cause of death in such countries, also affecting the mortality rate. In the UK which is at stage 4/5 of the Transition Model, healthcare is readily available and education about diseases such as AIDS is widely available, not only through schooling but through advertising such as on TV and the internet. Also food is more readily available whereas in Namibia food isn't as readily available and many people rely on the help of Aid Agencies and donations. Because people live longer in the UK (life expectancy 79 whereas in Namibia it is 50) the UK is definitely in Stage 4 where the death rate is 6/000.

This answer does provide some explanations for the high death rate in Stage 2 as opposed to Stage 4. It quotes some figures in support, which are of the right order even if they are not entirely accurate. The material on the UK is lacking, perhaps because time was running out. Nevertheless, the principles are explained briefly and in relation to the model.

Overall, this answer and the responses to the other two parts of the question should be enough to place this candidate on the A/B border.

Theme 2

Q3: Social structure and land uses

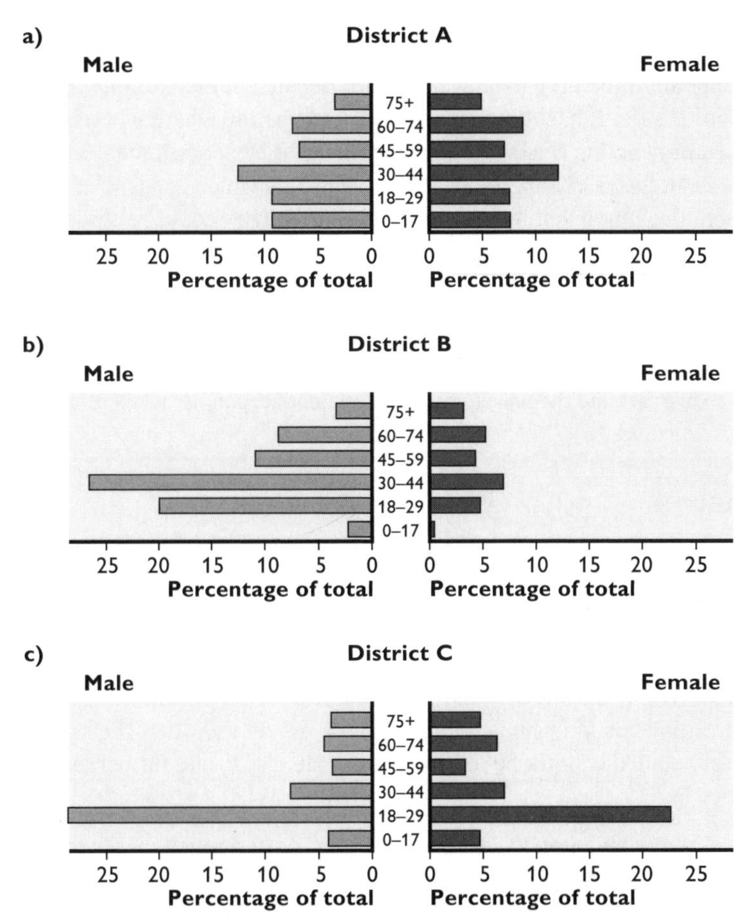

Figure 3 Age/sex pyramids for three districts of San Diego, USA

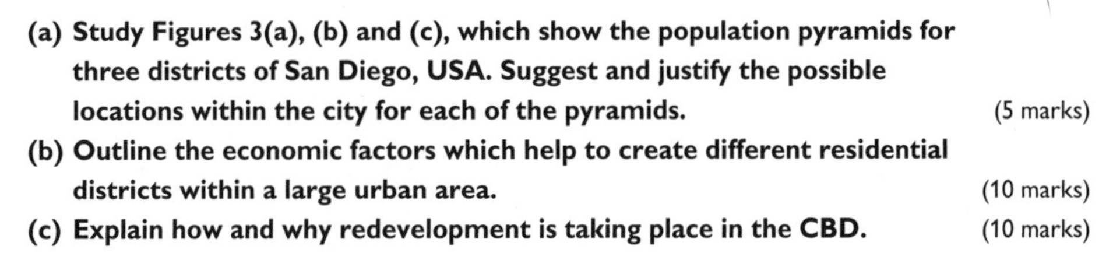

(a) Study Figures 3(a), (b) and (c), which show the population pyramids for three districts of San Diego, USA. Suggest and justify the possible locations within the city for each of the pyramids. (5 marks)

(b) Outline the economic factors which help to create different residential districts within a large urban area. (10 marks)

(c) Explain how and why redevelopment is taking place in the CBD. (10 marks)

C-grade answer

(a) District A is the inner city whereas district B is an area of immigration of males. There are many young people in District C which may be student housing.

The response is vague and lacks a good use of evidence hence it is probably a C grade.

A-grade answer

(a) District A appears to be an established suburb because it has a broad age range of population. District B has few females and could be anywhere where there are mainly males moving in. This could be an immigrant area or an area for the Gay population. District C is dominated by 18–29-year-olds which suggests it might be an area where the young first live after leaving home. They could be students.

This answer about District A is basically correct as it is a residential area in the suburbs. District B information is almost correct but it is not locating the pyramid although the reason is hinted at. It is close to the CBD and is an area housing young upwardly mobile males — not nest-builders as there are few children. District C is correct; it is an area around the university where studentification has taken place. This answer is worth an A.

C-grade answer

(b) The price of land is a factor as we see that most cheaper land is found further away from the CBD but houses might get more expensive with distance as rich people don't want to stay near the noise and other types of interference associated with the CBD.

The affordability of transport also plays an important role because if one cannot afford a car or pay bus fares he is going to locate nearer the CBD. The opposite is true because distance will not be a factor so people can locate further away from the city if they like.

This will see high density residential areas being formed much closer to the city centre in order to cut transport costs. The more wealthy people will locate in medium and low density areas that are found much further away from the city.

The other factor is rents of houses, you find that bigger houses are not nearer to the CBD because of the cost of land. So rents are higher out there than nearer the CBD. This creates different characteristics as flats are more in town than on the outside of town.

In the first and last paragraphs this essay does emphasise the role that land costs play (the term bid rent curve could have been used here). The second paragraph refers to costs of transport. The third paragraph could have been a conclusion. The essay only sees the city as inner versus outer and does not see any detailed division of districts resulting from migration, age and ability to pay (students), and even the role of gatekeepers (estate agents who promote areas to gain higher fees). The answer would be Level 2 because information and explanations are partial. There are no examples and the language is rather simplistic, which will keep the mark to the lower part of this band.

A-grade answer

(b) There are several economic factors that lead to different residential areas in the city. The bid rent curve says that rich people are willing to pay more for space and therefore the outer city tends to attract people with more money. In contrast poorer people can afford less space and are concentrated in terraced houses as in inner Swansea.

Proximity to good transport links will cause the price of property to rise and so those areas with good public transport or good, fast road links to the city centre will only be affordable by people with money.

The ability to get a loan for a mortgage will also lead to the areas of larger houses only attracting the middle classes. House agents can make places more middle class by only sending the middle classes to look at properties there.

Proximity to good schools will raise the price of property and so only those who can afford the houses will be able to buy near a school. This again segregates people. This happens in many towns such as Kingston.

The richest people often want to live very close to the city centre in regenerated areas of old warehouses or new apartment blocks such as the area around Cardiff Bay and London Docklands.

This answer has a range of economic factors some of which have examples added. The response could have a few more examples or improved detail to gain an even higher mark. Nevertheless it has A-grade understanding of the factors.

C-grade answer

(c) Redevelopment is taking place in the CBD for many reasons. First, the CBD is old and people want new buildings for shops and offices. Old shops are being knocked down and new ones built like Primark because people want cheap clothes because they cannot afford more. Sometimes you get big department stores that are new such as the John Lewis one in Cardiff. They had to knock down buildings first. City centres have lost shops to out of town shopping centres. Some shops have closed due to the credit crunch.

People like walking around shops and so the council has pedestrianised the main shopping street and sometimes they have covered the road to make it more pleasant to shop there.

There are now many offices in the CBD in skyscrapers. They are big because the companies that use them are big and need a lot of space. People journey to work from the suburbs and it is easier to reach the town centre. Nightclubs are always in the CBD because young people want nightclubs. There are many changes in the CBD.

question

This is a poorly composed answer yet it does have some knowledge within it. It does suggest that new shops are coming into the CBD and gives store names. It notes competition for the CBD but does not explain it. It has a section on pedestrianisation and some sentences on offices with some reasoning. Recreation is mentioned but without any reasoning in any depth. Only one place is mentioned although the whole answer is probably referring to Cardiff. The four changes mentioned would enable this response to creep to the C/D border. There is not enough explanation to gain more marks.

A-grade answer

(c) Redevelopment has taken place within the CBDs of many urban cities, including Birmingham in the last 20 years as a way of bringing back shoppers and residents. In the 1970s and 1980s Birmingham went through deindustrialisation and this led to large unemployment and derelict factories. The communities such as Aston also went into deprivation. This scares in the inner-city part investors off from the city and they chose to build new houses and shopping facilities as well as other services on the cheaper more accessible and aesthetically pleasing greenfield sites. This also leads to the decentralisation of retailing where large out-of-town shopping facilities were built away from the CBD. In the Birmingham conurbation, Merry Hill in Dudley, Marshall Lake in Shirley and Fortin Stourbridge are all examples of decentralisation of retailing. The Birmingham CBD was redeveloped with the flagship being the new Bull Ring shopping centre, employing over 8,000 people and costing over £500 million, the mail office in the zone of assimilation as well as the high quality apartments to attract residents to the CBD. The redevelopment and gentrification of Birmingham's CBD was an attempt to bring shoppers back to Birmingham, the Bull ring has 370 million visitors a year and invites wealthy people to live there. This will hopefully boost the status of the area, further encourage development and stop inner-city deprivation and decline.

This essay has identified how and why a selection of changes has taken place and provided some detail regarding retailing. It also refers to aspects of residence in the CBD. Its knowledge is worthy of an A grade, although the quality of language and lack of paragraphs does detract from the answer. It is very much on the lower border of an A.

Q4: Perceptions of rurality and CBDs

Richard Rowbottom

Figure 4(a) *Figure 4(b)*

(a) Use Figures 4(a) and (b) to suggest why perceptions of rural life are powerful pull factors in counter-urbanisation. (5 marks)

(b) Outline how perceptions of rurality have led to changes in the nature of rural settlements. (10 marks)

(c) Explain why redevelopment has taken place within central business districts in recent years. (10 marks)

A-grade answer

(a) Counter-urbanisation is when people move out of the city and into the countryside. As shown in Fig. 4(a), rural villages are seen as clean and tranquil places. Pavements have no litter and the vegetation is neat. Traffic levels are perceived to be low. People perceive rural life to include cottage-like houses with large gardens, a drive and maybe a garage as shown in Figure 4(b).

People perceive rural areas to be more civilised and suburban lifestyle includes having a local pub and familiar neighbours (Fig. 4(a)). As houses may have private drives (4b) people can park safely and easily in rural settlements and still be able to travel to work.

Figure 4(b) shows a pretty garden and 4(a) a pleasant clean community and people expect all rural lifestyles to be like this and it is an increased ideality.

This is a long response for 5 marks which uses the resources to make points as instructed in the question. It even recognises perception. Its made-up word ('ideality') at the end is not totally clear. Nevertheless it is an A grade answer.

C-grade answer

(a) The photos show a typical village with a quiet street running through it. There are many old buildings with many chimneys showing that they are old. People like moving to areas like this where there is a pub. This village will attract people with a lot of money, which is what counter-urbanisation is, rich people going to villages. They can commute by car to the big towns.

e The points made are generally along the right lines but they are never expressed that clearly. There is only one direct reference to the figure. The answer is too descriptive and lacking enough linkage to counter-urbanisation and perception to gain more than a C grade.

A-grade answer

(b) The perceptions of rurality have changed for different places. The case study we studied was Framlingham which had a dramatic population increase. From 1981 to 2001, the population increased as much as 42% from 2,190 to 3,114 and there was an estimate that it would increase up to 3,320 by 2006. House prices increased as much as three times in 17 years from £60,000 for a three bedroom semi-detached house, now the same house is worth £180,000. People perceive places like Framlingham to be the 'chocolate box' place where everything is perfect and top form but, in reality, the place may not be like that. There are problems associated with 35 hectares of open land lost from expansion of over 646 new houses. Also, traffic flows have increased to and from the village. From the current number of houses being created there has been an increased amount of flooding and overland flow from the new houses being built on greenfield sites and stopping the natural drainage from working. People perceive places to have nice weather but after everyone moves in their views change. Young people have been forced to move out as housing can be expensive and up to 10 times their average income.

e This response outlines the changes in the nature of the settlements quite well by using an example reasonably effectively. It lacks more information on perceptions of rurality other than the one reference to chocolate box villages. People perceive quiet, community, basic services nearby, and safety, as well as increasing value of property. Nevertheless, the answer would score an A because it uses a good example and it does outline the changes, one of which is environmental, thus linking changing physical environments to changing rural settlements.

C-grade answer

(b) Many people have the perception that rural areas such as Shropshire have relatively low house prices. However, due to so much demand for these rural houses the average house price in Shropshire has risen greatly in recent years.

Furthermore, many people also move their families out to a rural area with the perception that the area will be very quiet with very little traffic. However, due to more people owning 4X4s it is now a lot more busy in these areas. This is backed up by the fact that 55% of road related deaths now occur on rural roads. Therefore, the quiet and safe perception of rurality has now changed for the worse.

In addition, the perception of having lots of time and space to enjoy the countryside may also have changed due to many people living there having to commute miles to work in an urban area every day. This can also leave wives or children alone for long periods of time in a very isolated location. Therefore, the life with a strong sense of local community can also often be proved incorrect.

This answer has more about perceptions although it tends to outline these in general without supporting evidence from a real place. Just saying Shropshire does not illustrate a rural area *per se*. It would be better to look at a specific place such as Shawbury and use this village to illustrate the points made. Some of the ideas are vague, e.g. the points about traffic and even the implication that children are left on their own. This answer is likely to gain a C grade.

A-grade answer

(c) Inner-city areas are experiencing social deprivation due to closure of manufacturing industries. As the industries move out so do the more affluent people leaving the less affluent in the inner city. Cardiff Bay was once Dock land and a thriving port where vast amounts were exchanged.

As the coal industry declined so did a lot of the surrounding industries and they were forced to close down. This left an area of people without jobs who were relatively poor. The more affluent people filtered out towards the suburbs leaving Bute Town an area of working class people, who began to enter the cycle of poverty. The CBDC (Cardiff Bay Development Commission) began the regeneration of Cardiff bay.

The barrage was built to turn unattractive mud flats into an aesthetically pleasing environment that would encourage immigration and new development in the area. A whole complex of restaurants and bars were also built. The Millennium Centre was a huge development attracting a lot of tourists and became a main focus of Cardiff.

However, the accommodation was too expensive for the locals and they could not access all the facilities provided due to money problems. There is still a large area poverty stricken in Bute Town and surrounding areas as the regeneration has not helped break the cycle of poverty. The developments have not benefited who they were supposed to.

Also the lake has caused pollution due to algae and midges and flies are a huge problem. Many people protested against the development of the Barrage as it was disturbing the environment and local wildlife. The regeneration of Cardiff bay has however increased tourist numbers to the area and is an asset to Cardiff due to improvements in infrastructure.

The regeneration of Cardiff bay has not benefited who it was supposed to and has isolated locals and caused minor forms of segregation due to a divide into affluent and poor people. However it provides an interesting, modern, attractive environment and is a very popular entertainment district. It may have lost its architectural heritage and identity.

The answer responds well to the question's command and does evaluate. It has an introduction and a conclusion. It keeps to a single theme and discusses aspects of the regeneration. A sketch map would have taken this to another level. There is one small relevant addition after the conclusion which demonstrates that the student has thought about the answer.

Overall, this candidate would gain a good grade A for the three parts of the question.

C-grade answer

(c) In recent years CBDs (Central Business Districts) have been under threat from areas on the outskirts of cities. This is due to the growing number of out of town retail parks and business districts causing the CBD to lose business. Shops and services have also moved out of the centre of cities or towns because of the rising costs such as rent, of running a business there. This can cause the CBD to become almost derelict and crime and anti-social behaviour in the area can increase. It is because of these factors that the redevelopment of some CBDs, such as Reading (the Oracle Centre) and Bristol have taken place. In 2008, Cabot Circus was built in the centre of Bristol. Bristol's CBD was suffering because of out of town retail parks like Cribbs Causeway, so by building a stylish shopping area, people were encouraged back to the centre.

This is a rather short essay for a possible 10 marks. It develops one theme of competition from the city fringe and illustrates that with two examples and some reasons for the changes. However, the answer neglects other reasons for decline such as old buildings not adapted to modern life, traffic, access, and population shifts. Because it lacks breadth and lacks depth (small points such as mentioning whether a mall is indoors and climatised adds conviction) as well as detailed knowledge, this answer would gain a low C grade.

Research including fieldwork

Q5: Investigation into inner-city change

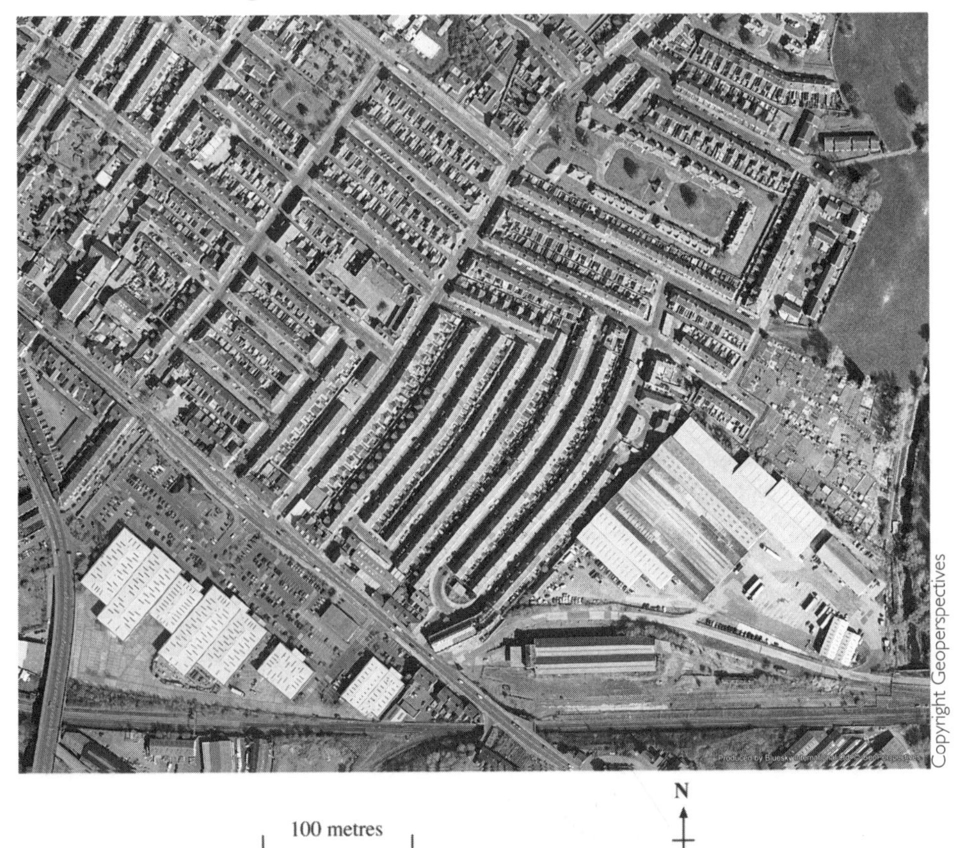

Figure 5 Vertical air photograph of part of an inner city in the UK

Study Figure 5, which was used in an investigation into inner-city change.

(a) Describe the variety of land uses seen on the photograph. (7 marks)

(b) Suggest and justify how you would plan a questionnaire survey to investigate ONE of the following aspects of inner-city change: (i) environmental, (ii) economic, (iii) social. (8 marks)

(c) Evaluate an investigation that you have undertaken into a changing human environment. You should clearly state the question that you have investigated. (10 marks)

A-grade answer

(a) Towards the NE there is a small park area for residents to walk through and use for leisure. Just to the south of this area is an area of allotments. In the south there are a large number of parked cars possibly 150 m from a station. However, they do seem to be near large buildings which are possibly used for retailing or business. To the SE there appears to be an old factory which once had rail access by the outline of its site. Just above it there is a newer factory with lorries parked. Much of the photo is covered by residential areas with an area of terraced nineteenth century housing for workers in the factories. Elsewhere the housing is semi-detached. There is a church on the west side.

The answer provides enough detail on a series of land uses to gain the full 7 marks. It uses compass points rather than top and bottom.

C-grade answer

(a) The majority of land uses in this photo seems to be housing as the centre of the photo and NW are housing. There is a mixture of housing in some kind of estate. To the very NE and E a green park stretches down the edge of the photo. In the SE there is a big factory and a road whereas in the SW there seems to be some sort of company.

The level of description lets this down. Several uses are identified but the language used becomes less geographical as it progresses. This response may get half marks for the broad typology of uses.

C-grade answer

(b) A questionnaire survey can only be planned by outlining simple and direct and closed questions piloted before performing the survey. Each of the environmental aspects such as litter, building quality, pollution or air quality should be mentioned in the survey. Possibly the people being surveyed can be asked to rate each of these aspects of change on a linear scale. Just as importantly though the sampling technique will need to be considered when constructing the survey. A random sample could be more suitable for example instead of a stratified or systematic method because this would prevent any biased results and make the questionnaire easier and more convenient to conduct. The number of people that will be surveyed also needs to be decided. I believe 50 persons will provide enough results for analysis but it still will not take too long to conduct.

This response tries to justify the type of sampling, the subject matter and the sample size. It also makes a comment about bias. It does not have to state the questions although it might have helped to illustrate good and poor questions. The suggestions and justification made in 10 minutes of writing time would merit a mark on the C/B margin.

D-grade answer

(b) A questionnaire to investigate economic inner-city change could be planned by firstly deciding what sort of questions you would ask and also who you would ask. You would need to plan because otherwise you might not get a target audience and also you would need to see what questions to ask that are relevant. Also you would need it spread out over an area and not condensed otherwise the results will lack validity. You would need to ask economic questions possibly to restaurants and ask about who they employ and how many and other details like this. Also how much the employees get paid. You might want to find out how much they get paid and if a wage affects living standards as well.

This is an example of a rambling answer which appears to say little but does have some vague ideas about planning. It does state, poorly, that you need to decide the number of questions, who to ask and where to ask. This answer is very weak. Consequently, it would struggle to gain marks.

A-grade answer

(c) I investigated how the environmental quality of the Central Business District in XXXXX is affected by pedestrian flows. I conducted a questionnaire, an environmental survey, a traffic count and a pedestrian count for various areas of the CBD. The main limitation of the investigation was down to timing. Considering that each area was only counted for two minutes, the results were not necessarily accurate. Furthermore, as myself and my assistants moved to different areas, we would be measuring the pedestrian counts and traffic counts at different points of the day. Overall there was about a 2 hour delay between the earliest and latest recordings, this would have affected flows. The questionnaire is subject to opinions, it may have been easy to conduct but the data was not completely foolproof. Likewise, counting traffic and pedestrians was an easy method but it is possible that human error would have influenced the recordings. Measuring and recording samples in a stratified approach was very useful. On the other hand I was able to locate pedestrianised and less pedestrianised areas and clearly show how the environmental quality was affected.

This answer tells us what the study was and does note, in passing, that it was an exercise that used fellow students. There is nothing wrong with group work. A range of methods of data collection are described and there is some evaluation of the issues of data collection. The answer does not go on to evaluate the whole exercise although the last sentence implies a degree of success. This response is on the A/B boundary.

Overall, this first candidate would gain a grade A mark because the excellent performance in (a) and the very good performance in (c) would make up for the slightly less strong performance in (b) (the C/B grade).

question

A-grade answer

(c) The question that I investigated was 'What factors affect environmental quality in XXXXX?' We used quite a few methods, which we thought might help us investigate our question on environmental quality. We did pedestrian counts, car counts, questionnaires and bi-polar analysis. The strength of using a questionnaire is that you can get good qualitative data which you can then compare to other data. Also you get lots of opinions which you can compare. A weakness is that people may exaggerate and give socially desirable answers which will affect validity. A strength with pedestrian counts is that you are able to see how populated an area is and how that correlated to people's opinions in the questionnaires. A weakness is that the pedestrian counts can only give you an idea of environmental quality if you have something to correlate with, e.g. pedestrian count and litter count. Another method we used was bi-polar analysis of an area — this is when we had to evaluate an area in terms of ugliness/prettiness or pollution or overcrowding, usually done on a 5 point scale. A strength is you can get an idea of an area and its qualities, but a weakness is that you are only getting the one person's opinion and that is whoever is filling it . You improve on this by getting more opinions from others. The sampling we used for the questionnaire was when we went into an area and just chose whoever was nearest to us — convenience sampling. This was positive because it was easy to do but a weakness was that it could be subjected to bias from the person asking the questions. Improve this by systematic sampling.

e This scores more than the other candidate because it evaluates by looking at the good and bad points of the method used on the same survey as the first candidate. It also examines an extra technique, bipolar analysis. The last sentence suggests that the student ran out of time. Do use the time guide on page 5 of the Introduction and don't rush this last 10 mark answer. If complete, this answer would receive an even higher A mark than the first candidate's answer.

Overall, the strong response to part (c) has raised the second candidate's final grade for the question to a C. Do always bear in mind that a strong answer to a part can help you get a good mark. Do not be put off by the first parts of a question.